CROSSBILL GUIDES

The nature guide to the

Cévennes
AND GRANDS CAUSSES

FRANCE

The Nature Guide to the Cévennes and Grands Causses

Initiative, text and research: Dirk Hilbers, Paul Knapp
Additional text, Suzie Coulton, Albert Vliegenthart (insects)
Additional research: Kim Lotterman, Albert Vliegenthart
Editing: John Cantelo, Brian Clews, Jack Folkers, Cees Hilbers, Riet Hilbers,
    Kim Lotterman, Henk Strijbosch (reptiles and amphibians)
Illustrations and maps: Horst Wolter
Type and image setting: Gert Jan Bosgra

Special thanks to Jan van der Straaten / Saxifraga Foundation

Print: Ponsen en Looijen, Wageningen

ISBN 978 90 5011 279 6

© Crossbill Guides Foundation, Arnhem, The Netherlands

This book is published in association with WILD Guides, KNNV Publishing
and the Saxifraga Foundation.

www.crossbillguides.org
www.wildguides.co.uk
www.knnvpublishing.nl

WILD Guides

KNNV Publishing

KNNV vereniging voor veldbiologie

SAXIFRAGA foundation

CROSSBILL
GUIDES
FOUNDATION

4

About this guide

This guide is meant for all those who enjoy being in and learning about nature, whether you already know all about it or not. It is set up a little differently from most guides. We focus on explaining the natural and ecological features of an area rather than merely describing the site. We choose this approach because the nature of an area is more interesting, enjoyable and valuable when seen in the context of its complex relationships. The interplay of different species with each other and with their environment is simply mind-blowing. The clever tricks and gimmicks that are put to use to beat life's challenges are as fascinating as they are countless.

Take our namesake the Crossbill: at first glance it's just a big finch with an awkward bill. But there is more to the Crossbill than meets the eye. This bill is beautifully adapted for life in coniferous forests. It is used like a scissor to cut open pinecones and eat the seeds that are unobtainable for other birds. In the Scandinavian countries where pine and spruce take up the greater part of the forests, several Crossbill species have each managed to answer two of life's most pressing questions: how to get food and avoid direct competition. By evolving crossed bills, each differing subtly, they have secured a monopoly of the seeds produced by cones of varying sizes. So complex is this relationship that scientists are still debating exactly how many different species of Crossbill actually exist. Now this should heighten the appreciation of what at first glance was merely a plumb red bird with a beak that doesn't close properly. Once its interrelationships are seen, nature comes alive, wherever you are.

To some, impressed by the "virtual" familiarity that television has granted to the wilderness of the Amazon, the vastness of the Serengeti or the sublimity of Yellowstone, European nature may seem a puny surrogate, good merely for the casual stroll. In short, the argument seems to be that if you haven't seen some impressive predator, be it a Jaguar, Lion or Grizzly Bear, then you haven't seen the "real thing". Nonsense, of course.

But where to go? And how? What is there to see? That is where this guide comes in. We describe the how, the why, the when, the where and the how come of Europe's most beautiful areas. In clear and accessible language, we explain the nature of the Cévennes and Grands Causses and refer extensively to routes where the area's features can be observed best. We try to make the Cévennes and Grands Causses come alive. We hope that we succeed.

How to use this guide

This guidebook contains a descriptive and a practical section.

The descriptive part comes first and gives you insight into the most striking and interesting natural features of the area. It provides an understanding of what you will see when you go out exploring. The descriptive part consists of a landscape section (marked with a red bar), describing the habitats, the history and the landscape in general, and of a flora and fauna section (marked with a green bar), which discusses the plants and animals that occur in the region.

The second part offers the practical information (marked with a purple bar). A series of routes (walks and car drives) is carefully selected to give you a good flavour of all the habitats, flora and fauna that the Cévennes and Grands Causses has to offer. At the start of each route description, a number of icons give a quick overview of the characteristics of each route. These icons are explained in the margin of this page. The final part of the book (marked with blue squares) provides some basic tourist information and some tips on finding plants, birds and other animals.

There is no need to read the book from cover to cover. Instead, each small chapter stands on its own and refers to the routes most suitable for viewing the particular features described in it. Conversely, descriptions of each route refer to the chapters that explain more in depth the most typical features that can be seen along the way.

We have tried to keep the number of technical terms to a minimum. If using one is unavoidable, we explain it in the glossary at the end of the guide. There we have also included a list of all the mentioned plant and animal species, with their scientific names and translations into German and Dutch.

Some species names have an asterix (*) following them. This indicates that there is no official English name for this species and that we have taken the liberty of coining one. For the sake of readability we have decided to translate the scientific name, or, when this made no sense, we gave a name that best describes the species' appearance or distribution. Please note that we do not want to claim these as the official names. We merely want to make the text easier to follow for those not familiar with scientific names. When a new vernacular name was invented, we've also added the scientific name.

An overview of the area described in this book is given on the map on page 12. For your convenience we have also turned the inner side of the back flap into a map of the area indicating all the described routes. Descriptions in the explanatory text refer to these routes.

walking route

car route

beautiful scenery

interesting geology

interesting flora

interesting invertebrate life

interesting reptile and amphibian life

interesting birdlife

visualising the ecological contexts described in this guide

Table of contents

Landscape   9
  Geographical overview   11
  Grands Causses at a glance   14
  The Central Cévennes at a glance   18
  Geology and Climate   22
  Habitats   28
  Mediterranean scrub   30
  Forests   34
  Subalpine meadows, heath and bogs   41
  Rivers and river gorges   45
  Causses - dry plateaux and karst   49
  History   56
  Nature conservation   67

Flora and Fauna   71
  Flora   74
  Mammals   98
  Birds   101
  Reptiles and amphibians   114
  Insects and other invertebrates   119

Practical Part   135
  Route 1: Touring the northern Cévennes   137
  Route 2: The Causse and the gorges   142
  Route 3: The chestnut groves of the eastern Cévennes   148
  Route 4: Causse du Larzac   153
  Route 5: To the Mont Aigoual   160
  Route 6: Causse Blandas   167
  Route 7: The mires of Mont Lozère   170
  Route 8: The summit of the Mont Lozère   174
  Route 9: Mas Camargues   177
  Route 10: The Menhirs of les Bondons   180
  Route 11: Bois de Paiolive   183
  Route 12: Le Can de l'Hospitalet   186
  Route 13: The Chestnut forests of les Roquettes   188
  Route 14: Walking the Feather Grass steppes   191
  Route 15: Nimes-le-Vieux   194

Route 16: Where the Tarn and Jonte Gorges meet   197
Route 17: The schist Cévennes   201
Route 18: Orchids in the Cernon valley   204
Route 19: Gorge de la Vis – Cirque de Navacelles   208
Interesting sites and other extras   211

Tourist information and observation tips   **215**

Glossary   **230**

Picture and illustration credits   **231**

Species list and translation   **232**

List of Text boxes
Caves and Karst; water as a sculptor   24
Lavognes –Lifeblood of the Causses   52
Przewalski Horses   69
Thistle on the door   85
The decline and successful recovery of the Griffon Vulture   104

LANDSCAPE

Cévennes

The Cévennes and Grands Causses in the south of France are a naturalist's heaven. There are few places in Europe that match its diversity of interesting sites, landscapes and species. Each landscape has an appeal of its own. You will find dazzlingly deep gorges and large, mossy woodlands. You can walk in gentle hills among colourful and sweetly scented meadows, but also trek dramatic and empty, windswept plateaux with large boulders scattered over scanty heathlands. However different in character these trails may be, all have as a common factor, the splendid and unspoilt landscape of the area.

This landscape is the backdrop of an exceptionally diverse flora and fauna. From mammals to moths and from birds to butterflies, the variety of wildlife is enormous. The flora, in particular, is outstanding, with many species of orchids, lilies and other attractive wildflowers. The insect life, notably the butterflies, is enchanting and there is also a good range of reptiles and birds. The interesting plants and animals are not confined to a certain spot, but seem to be everywhere. The Cévennes is one of those regions that can leave you torn between trying to reach the intended goal of a daytrip or lingering a little longer en route to identify that particular orchid or photographing that stunning Green Lizard. It is easily possible to spend a full day on the first kilometre of any trail.

The Cévennes also has a long history of human habitation. The villages and towns in the region are old and picturesque. During the last century an exodus from the countryside took place, leaving many of the hamlets and farmsteads depopulated with groves and terraces abandoned. Every mountain and valley has its own story to tell and the wandering rambler will come across many unexpected and often beautiful remains of past cultivation. From the ecologist's perspective, the Cévennes are equally mesmerizing. The region's position on the edge of the Mediterranean and several other climate zones, plus the complex geology, creates a fascinating ecology which waits to be unravelled.

The Gorge de La Jonte is one of the most spectacular gorges of the Cévennes and Grands Causses (route 16).

10

The diverse landscape (bottom) of the Cévennes attracts a wide range of species, including Red-backed Shrikes (left) and Scarce Swallowtails (right), both of which are common sights in the area.

After a long day in the field, the Cévennes offers plenty of scope for relaxation. The many hours of sunshine provide the opportunity, the pretty streamside shingle-banks the location and the fine wine the excuse to just sit and relax as you digest the kaleidoscope of impressions that a day in the Cévennes offers. It is indeed a naturalist's heaven.

Geographical overview

The Cévennes and the Grands Causses together form the south-eastern part of the 'Massif Central' in southern France. To the south, the mountains drop down to the Mediterranean coastal plain near Montpellier. To the east the mountains give way to the river valley of the Rhone and, travelling west of the Grands Causses, you enter the lovely region of the Dordogne. To the north the boundary of the Cévennes is less clear, as the region gives way to similar mountain ranges. Travelling to the north-east of the Cévennes one finds oneself suddenly in the Ardêche, a limestone mountain range with similarities to the Cévennes. The predominantly granite Auvergne (Aubrac and Vivarais mountains) is situated northwest of the Cévennes.

The Cévennes and Grands Causses covered by this guidebook contain three distinctive regions, the Grands Causses in the west, the high Cévennes in the centre, largely coinciding with the Cévennes National Park, and the Cevenol Valleys in the east.

Grands Causses

The Grands Causses are rugged and empty upland limestone plateaux, in-tersected by spectacular, forest clad gorges. These are cut out by the rivers Lot, Tarn (with its famous Gorge du Tarn), Jonte and Dourbie. The plateaux that separate these rivers are the Causse de Sauveterre (between Lot and Tarn), the Causse de Méjean (between Tarn and Jonte), the Causse Noir (between Jonte and Dourbie) and the Causse du Larzac (between Dour-bie and the Mediterranean plain). Several smaller Causses flank these big four. The large Parc Naturel regional des Grands Causses encompasses most of the region.

The landscape of the limestone tablelands (Causses) resembles that of the African steppes and Savan-nahs.

Overview of the
Cévennes and
Grands Causses.

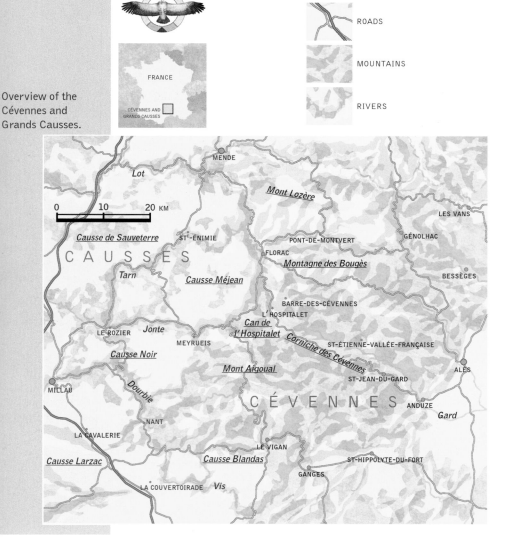

The tablelands are open, sparsely populated areas, but the river gorges harbour some villages and towns, the largest being Millau (with its famous new bridge), Le Rozier, Meyrueis and St. Enémie. The town of Florac, the most centrally situated town of the Cévennes, lies in between the Causse Méjean, and the central ridge of the Cévennes.

Further west, outside the area covered by this guide, lies the main town of this area, St. Affrique.

The Cévennes National Park

The backbone of the Cévennes is a ridge that runs from the Mont Lozère (1,699 m) south-south-east down towards the Mont Aigoual (1,565 m), the two highest peaks of the area. This ridge is the core of the Cévennes National Park. The Cévennes mountains are higher than the Causses and consist largely of schist and granite bedrock, thereby boasting a completely different, and much greener, landscape than the dry Causses. There are only a few larger municipalities hugging the central Cévennes ridge, the most important ones being Florac on the western side, Genolhac in the north-east and Le Vigan in the south.

The Cévenol ridges and valleys

East of the central Cévennes, numerous smaller ridges run south-east in the direction of the Rhone river valley. This lovely mountain landscape is heavily wooded, with crystal-clear mountain streams running through its valleys. Here, larger villages are few, but hamlets are numerous and are connected by a dense network of narrow country roads. It is the heartland of the historical Cévennes, with its chestnut collectors, its history of Camisard rebels (see page 59) and its silk production (see page 63).

The larger towns of the Cévennes are located at the base of the actual

The cool granite slopes of the Mont Lozère remind of Ireland.

The ridges and valleys of the 'proper' Cévennes are covered in forest and heathlands.

mountains in the west and south of the area. From north to south these are Les Vans, St. Ambroix, La Grand-Combe, Alès (the only true city in the area), Anduze, St. Jean-du-Gard, Ganges and Le Vigan. Here, the low hills at the eastern and southern rim of the massif consist of limestone again and have a distinct Mediterranean character.

The Grands Causses at a glance

The Causses are a series of rolling limestone uplands, divided by deep river gorges. The Causse landscape somewhat resembles, on a vast scale, cracked clay: slabs of relatively smooth, undulating land, divided by sudden deep crevices. The gorges and the plateaux are two completely different worlds. In the gorges there is water and lush vegetation. Shrubs, trees and a wide array of wildflowers appear from the cracks and the not-too-steep sections of the gorges. Villages are crammed into the few places where the ground is relatively flat. Old stone houses have been constructed on rocky promontories along the river or seemingly glued onto impossibly vertiginous rocky slopes. The gorges differ somewhat in appearance. The spectacular Gorge du Tarn is a popular tourist destination for canoeists and rafters, but the Gorge

The Causse seems to be torn off at the edges, where very sudden, deep gorges separate them from the opposite Causse.

de la Jonte is more wild and its river too shallow for canoes. The Dourbie by contrast is not as rocky, and clad in beautiful oak woodland.

When you snake up one of the minor roads to the plateau, a completely different landscape unfolds. On the Causse you move through a wide, empty landscape with rocky grasslands, shrubs and sudden rock pillars. The Causses – in particular the Causse Méjean – are the least populated areas of France. A handful of hamlets fashioned from massive blocks of limestone, shelter the few shepherds and sheep-cheese farmers who still subsist here from the relentless elements. Winters are cold and windy, with snowfall sometimes between the months of January and March. Summers, in contrast, are hot with a fierce sunshine.

The word Causse is an old Occitan word, derived from the Latin Calx: chalk or limestone. The name couldn't be more accurate, because the Causses consist entirely of porous limestone with rain-water seeping away immediately into the limestone soil, so natural surface water is almost entirely absent. Hence, the plateaux have a dry and rugged appearance with a vegetation somewhat intermediate between Southern England's chalk hills, Mediterranean scrubland and the Ukrainian steppes. The Causses are primarily used for grazing sheep (The famous Roquefort Cheese comes from the Causses). To water the sheep, shepherds constructed large artificial pools, called *Lavognes*, which dot the dry and empty landscape.

However, the Causses are not everywhere this steppe-like. Many parts have been planted with trees, both for wood production and to prevent soil erosion. Moreover, the Causses differ between themselves as well. Below we present a

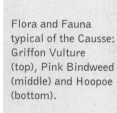

Flora and Fauna typical of the Causse: Griffon Vulture (top), Pink Bindweed (middle) and Hoopoe (bottom).

brief description of each of them including references to the routes (page 137 onwards) by which they may be explored.

Causse Sauveterre (route 1)

The Causse de Sauveterre is the northernmost and the second largest of the Grands Causses. It has all the interesting Causse features, such as Feather Grass steppes, scrublands of Box and Snowy Mespil, little stone hamlets and pine woodlands, but it also harbours a relatively large percentage of arable fields and therefore has a less pronounced steppe character. As such, the Sauveterre is the less spectacular of the five Causses.

Causse Méjean (routes 2, 14, 15 and 16)

The Causse Méjean is the most famous of the Causses and arguably the most beautiful. Bordered by the Gorge du Tarn and the Gorge de la Jonte it is the most distinct plateau with sheer cliffs all along its edges (except in the south-east where the Col de Perjuret connects it to the central Cévennes). This section, which is part of the Cévennes National Park, is the most open and deserted part of the entire Causses region. Here, one is forgiven for having the impression of being on an Asian steppe instead of France. No wonder then, that it is here that the National Park chose to breed the endangered wild Przewalski Horses for reintroduction into their homeland, Mongolia. Together with the Causse Blandas (see next page), the Causse Méjean is the most interesting for birdwatching.

Grasshoppers, like this Saddle-back Bush-cricket, do well on the steppes of the Causses.

Causse Noir

The Causse Noir is much more wooded than the others. It is named Noir (black) because of the many Austrian Pines that grow here. Therefore, the landscape does not have the steppe-like atmosphere that the others have. Instead, it has the pleasing, quiet character of woodlands, alternated with areas of flowery limestone grasslands and meadows. The flora of the Noir is superb, matching that of the Causse Méjean and Larzac.

Causse Larzac (routes 4 and 18)

The Causse du Larzac is the largest of the Causses, and together with the Causse Blandas, the

The limestone up-land plateaux (top), or Causses, are very rich in plant, bird and butterfly species.

A Bee Fly on Bladder Senna in the Gorge du Tarn (bottom).

most southerly. It is, quite literally, a lookout over the Mediterranean. The Causse du Larzac is a vast open territory, in several ways similar to the Causse Méjean, but less visited. Larzac is dissect-ed by the A9 Motorway, which leaves the Causse in a northern direction over the Millau Viaduct. The Causse du Larzac is probably the most in-teresting for searching wildflowers and butter-flies, because it harbours many Mediterranean species. There is also a species of orchid growing here that can be found nowhere else in world but on Larzac: the Aveyron Orchid* (p. 206).

Causse Blandas (route 6 and 19)

The Causse Blandas is the smallest of the five Causses discussed here. It is a sort of sub-Causse of the Larzac and is equally influenced by the Mediterranean. Some birds and butterflies that you won't find anywhere else in the Cévennes, occur here. Its southern border is the Gorge de la Vis, which has a fully Mediterranean vegetation and fauna.

The fresh meadows on the Lozère support various species of Pinks, including this Sequier's Pink* *(Dianthus seguieri)*.

The Central Cévennes at a glance

The ridge of the central Cévennes is the highest of the region. It lies entirely within the National Park boundaries. The landscape of the central Cévennes is the most diverse, with a mixture of woodland, scrub, meadows, karst, cliffs, chestnut groves, mountain heathlands and bogs. The central Cévennes includes elements of the Causse and of the Schist landscape of the eastern Cévennes, but in addition there are two large granite expanses, which form the two highest mountains of the area, the Mont Lozère and the Mont Aigoual.

Mont Lozère (routes 1, 7, 8 and 9)

Traversing the Mont Lozère, you might think that you are in Ireland. The fresh green hills, the mires, the little stonewalls, the broom fields with rounded granite boulders all remind of the open Atlantic coastlands of Ireland. This is the landscape that typifies the highest (and wettest!) parts of the Cévennes. The Mont Lozère (1,699 m) catches 1,600 mm of rain each year. Bogs soak up the water like a sponge and then release to the land below. Countless little streamlets come together to form the source of the river Tarn and countless tributaries flow into the River Lot in the north. Looking at the green pastures it is no wonder that the flocks of sheep of the Mediterranean plain move towards the Lozère in the summer months (see page 27).

From an ecological perspective, the Mont Lozère and Mont Aigoual are part of the Atlantic Mountains: on average cool, windswept and wet landscapes with heathlands and mires.

The Lozère is not a jagged mountain like the Alps or the Pyrenees. On the contrary, from a distance the Mont Lozère resembles a monk's tonsure. The rounded high slopes are treeless, surrounded by a belt of mountain forest of beech and spruce trees at lower altitudes.

Montagne de Bougès and Can de l'Hospitalet (routes 3 and 12)

Between Mont Lozère in the north and Mont Aigoual in the south lie the Montagnes de Bougès and the Can de l'Hospitalet. The first is a largely deserted and heavily forested mountain range, fairly similar to the schist Cévennes in the east, but higher.

South of the Bougès lies the Can de l'Hospitalet, which is like a small Causse. The Can is a dry, limestone plateau with rocky steppes, scrub and small woodlands. It differs from the 'true' Causses for having a more complex geology which gives it a somewhat different vegetation.

Mont Aigoual (route 5)

The Mont Aigoual is, at 1,565 m, the watchtower over the Mediterranean. It is in many ways a small Mont Lozère: a rounded mountain with granite bedrock, only the Mont Aigoual has a particularly nasty microclimate. More rain falls here than on the Lozère. It comes down in fierce autumn and winter deluges and the exact amount of precipitation varies strongly from year to year. The extent of mountain heathland on the Aigoual is much less than on the Lozère. Instead, the Aigoual has a lot of woodland. Although much of it is rather uninspiring spruce plantation, there are some beautiful Beech woods.

Mature forests of Spruce and Beech grow at an altitude of around 12 - 1400 metres, close to the summits of the Mont Aigoual and Lozère.

The Schist Cévennes at a glance (routes 3, 13 and 17)

The area east of the National Park is rippled with ridges and incised by valleys that run from the high Cévennes in east-southeast direction towards the Rhone Valley. This area is rugged, sparsely populated and covered with large expanses of pine and chestnut forests, alternating with warm heathlands and scrubland in the valleys. Dull and ponderous schist rocks protrude from the vegetation bearing testimony to the meagre soil of this region. In contrast to the limestone soils, schist is not very porous, so instead of being absorbed, water flows off the mountains in creeks and streams, which have carved out V-shaped valleys in the bedrock. Notwithstanding the presence of water, the poor soil is only marginally suited for agriculture. The only crop that flourishes in this area is the Sweet Chestnut. Indeed, upon closer inspection, many of the chestnut woods turn out to be old groves (see page 62). Dotted across the hillsides, amidst the foliage of these woods, are the old farmsteads of the Cevenol people that lived off the chestnut trees.

From a historical perspective, this remote area, just outside the National park proper, is the true Cévennes. The whole area exudes an atmosphere of a rural life with a long history and rich in tradition, but also one of great hardship. A visit to the Cevenol ridges and valleys is inevitably not only for the naturalist, but also for those interested in the history and ethnography of the Cévennes.

The eastern and southern rim (route 11)

The Cevenol ridges lose altitude towards the east and the south, where they eventually give way to the floodplain of the River Rhone and the Mediterranean. On the edge of the Cévennes the hamlets have become villages and towns. The solitude of the chestnut hills have made way for lively settlements such as Alès, St. Jean du Gard, St Ambroix, Le Vigan and others. These towns, still very much part of the Cévennes, are the gateways to the hinterlands.

The vegetation in the hot lowlands is typical Mediterranean scrubland, with seas of wildflowers in the mild spring and the loud rasping sounds of Cicada in scorching summers. For the naturalist, there is still a lot to discover here, particularly in spring.

Dry heathlands occur on schist soils. In good years, they are purple throughout the summer season.

22

Geology and Climate

Superb examples of karst features are found on the Causses (routes 2, 4, 14 and 15) and in the Bois de Païolive (route 11). You traverse granite boulder landscapes (ruines in French) on routes 1 and 7. The Ecomusée of le Pont de Montvert has a small exhibition on the Cévennes geology (in French). On page 214 you'll find an overview of caves, karst and other landscape features associated with limestone.

The key to the natural richness of the Cévennes lies in its climate and its complex geology. Three completely different types of bedrocks are found in the region: limestone, granite and schist. These different bedrocks have shaped the landscape in such a fundamental way that you can't really speak of THE Cévennes, rather of the schist, granite and limestone

Simplified map of the geology of the Cévennes and Grands Causses.

Facing page: the three different rock types found in the Cévennes. Limestone (top), Granite (middle) and Schist (bottom).

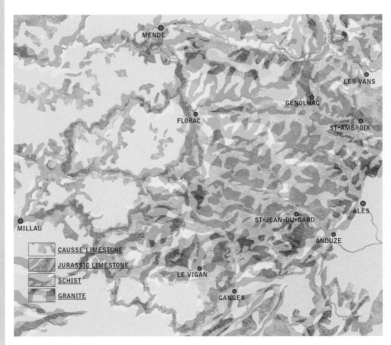

MENDE

LES VANS

GÉNOLHAC

FLORAC

ST-AMBROIX

ALÈS

ST-JEAN-DU-GARD

MILLAU

ANDUZE

CAUSSE LIMESTONE

JURASSIC LIMESTONE

SCHIST

GRANITE

LE VIGAN

GANGES

Cévennes. The bedrock shapes not only the landscape and its vegetation but also determines its flora and fauna and even the cultural history. Therefore you will find that this book is laced with references to these soil types. It also dominated our choice in selecting the routes, balancing them out over the different geological regions. For instance, the introductory car routes (routes 1 to 4) we organized in such a way that they capture two and, where possible, all three geological worlds in a single trip.

Geological history

The bedrock of the Cévennes was shaped in two different eras. The schist and granite was deposited early in geological history. Later, parts of the terrain were overlain with limestone.

The oldest bedrock dates from the late Carboniferous age, some 290 million years ago. It is part of the Hercynian mountain range, a huge range that came into being during an ancient collision of tectonic plates. Hercynian massifs are among the older mountains in the world. They were created long before the 'Alpine' mountains of the Pyrenees, the Picos de Europa, the Carpathians and, of course, the Alps themselves.

Most of the Massif Central is of Hercynian origin and is part of a now disjointed range that comprises the Vosges in Northeast France, the Ardennes in Belgium,

the Black Forest in Germany, Brittany in the west of France and parts of the southern UK. Towards the southwest, the Hercynian mountains continue in western Iberia. A twin chain was formed in the same process, the Apalachian mountains, now situated thousands of kilometres west on the

Caves and Karst; water as a sculptor

There are many caves in the Cévennes famed for their beauty. A quick glance on the map on page 213 however, reveals that all these caves are situated on the Causses and in the southern and eastern periphery of the region. The heartland of the Cévennes has none.

Most caves are present in limestone regions because limestone has a unique feature; it dissolves in water. Rainwater is naturally slightly acidic, causing alkaline (basic) rock like limestone to dissolve. In contrast, schist, which is acidic, does not dissolve. Schist erodes easily too, but that is due to a mechanical process in which gritty particles suspended in water physically erode the rock's surface. With limestone the weathering is a chemical process in which water literally eats its way through the bedrock. This is why the rivers in limestone regions create such spectacular gorges.

But this alone doesn't explain the dramatic features associated with this type of rock. Limestone does not have a uniform structure. Limestone is comprised of particles of dead marine micro organisms that sunk to the sea bottom. Since the species composition and abundance of micro organisms varied from place to place, the resulting limestone varies in its resistance to chemical erosion. Erosion creates little dimples or cracks, in which water accumulates. Since limestone weathering is a chemical process, the limestone is eaten away, particularly the "soft" stone, even when the water isn't flowing. Gradually, over time, the rain water sinks deeper and deeper in the limestone plateaux following the softer rocks, sometimes leaving clefts that run hundreds of metres deep, in some places narrowing to a thin crack where the bedrock is harder and then opening up again where a large part of the mountain is softer. Huge empty holes were created where there once was a concentration of soft limestone, like the Aven Armand cave.

In these caves, the water-limestone interaction has created even more spectacular scenery. Water in the droplets hanging from the cave ceiling slowly evaporates and the limestone solute recrystallizes on the ceiling. Over the centuries the crystalline structures grow into large dagger-like forms, the so-called stalactites. A similar process takes place when the droplets fall down on the bottom of the cave. Then upward growing structures, stalagmites, are the result. If stalactites and stalagmites meet, they may merge into magnificent pillar-like shapes. The spectacular colours caused by iron and sulphur salts add the finishing touch.

But caves are just one category of sculptures made by the water. In a similar fashion the soil above the ground weathers away to stand above the surface a haphazard collection of rock pillars that erode less quickly. This karst landscape, as it is called, can take the form of wondrous petrified "cities", like Nîmes-le-Vieux (route 15) and Montpellier-le-Vieux. Other superb karst features are the stunning Bois de Païolive (route 11) and various rock formations on the Causse du Larzac (route 4).

The sea bottom that provided the 'raw material' for the present-day limestone didn't consist of calcareous skeletons of micro organisms alone. Sand and clay were mixed in with it. Rainwater that dissolves the limestone carries this clay with it. It accumulates in the depressions in the land, called dolines, which become much richer in nutrients. These circular patches are the only places on the Causses where crops can be grown and these sites are readily visible in the landscape.

Apart from these dolines, and the occasional man-made lavogne (see page 52) the Causses are dry. There is no surface water because it sinks directly into the ground. It is incredible but true: on the Causse, one can die of thirst in a dry and desolate landscape, while knowing that at that very moment, the surrounding landscape is being shaped by water.

The most important geological features associated with limestone.

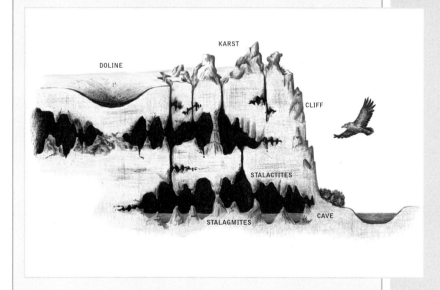

other side of the Atlantic.

Hercynian rocks are metamorphic rocks, meaning that they were formed when sediments were transformed under extreme pressures and temperatures. In the Cévennes, it is mainly the schists that formed from old sand and clay sediments of an ancient seabed. These were turned into rock by the pressures which coincided with formation of the mountains. Deeper below the earth's surface, the bedrock melted and was forced up through the schist layer where it slowly cooled to become granite. With the continued folding of the Hercynian mountains, this granite rose to the surface at several locations which now form high points like the Mont Lozère and Mont Aigoual.

Much later, about 30 million years ago, the water level rose and the Thethys Sea (of which the present-day Mediterranean is a remnant), flooded inland and submerged what is now much of southern France. The Cévennes in those days were probably a group of islands in a tropical ocean. Over the millennia the ancient seabed around these proto-Cévennes islands was covered by river sediments, by shells and by many millions of chalk skeletons of marine micro-organisms that lived in the warm, shallow waters. They created a thick layer that, over the ages, turned into limestone.

The tectonic clash that created the Alps and Pyrenees also pushed up the Massif Central. But in contrast to the Pyrenees and Alps, which folded the strata into dramatic loops and curves, the limestones of the Massif Central rose up without much disturbance to their original flat, layered bedrock. The Cévennes limestone regions, therefore, kept their original form and relatively flat surface. These later became incised by rivers which created the present tableland structure of the Causses.

Geology determines the land use

The schist region, which historically became known as the Cévennes, was poor, remote, sparsely populated and, above all, known as the region of endless chestnut groves. All these characteristics are a direct result of the infertile and easily eroding schist. Schists can support little but heathlands and chestnuts (see page 62). The numerous rivers that run down from the high Cévennes created the V-shaped river valleys that are a typical feature of the Schist Cévennes.

As already noted the granite regions are largely situated in the highest parts of the Cévennes, the Mont Lozère and Mont Aigoual. Again they form a poor soil, and worsened by a harsh climate with extreme winds and rain,

the granite uplands are inhospitable up to the point that permanent habitation is almost impossible. However, the conditions did provide a good summer pasture for sheep. This reflects in a history of transhumance and shepherding that dominates the history of the Mont Lozère.

The limestone regions of the Grands Causses are very different from the 'proper' Cévennes. The porous limestone easily erodes, which results in deep valleys, where arable farming is possible on terraces. However, on the high tablelands with a permanent lack of fresh water, growing crops is only possible in fertile depressions in the landscape, known as dolines (see text box on page 24). The rest of the land was suitable only for sheep grazing. Further west, on the lower Causses, the land becomes more fertile. This part became famous for its sheep's cheese and wealthy due to the economic activity encouraged by the arrival of the Templars (see page 60). The sheep's cheese was also produced in the high Causses.

Transhumance is a typical Mediterranean custom. The hot lowlands provide food only during the winter, whereas the mountains only have grass in summer. Hence, the flocks are moved twice a year between the plain and the mountain.

Habitats

A habitat, from the Latin *habitare* – to live, refers to a natural unit of plants and animals that live together. Such a unit is characterised by its own unique combination of geological structure, flora and fauna, ecological rules and appearance. So obviously, the more habitats that are distinguishable in an area, the more diverse and interesting it is.

In the Cévennes and the Grands Causses no less than 55 habitat types are found. This incredibly high number is a result of the complex geology and diverse climatic conditions in the region.

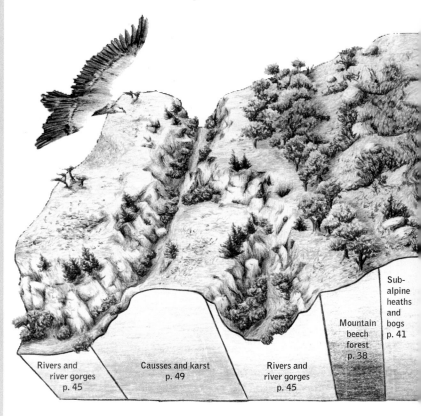

Overview of the habitats of the Cévennes and Grands Causses.

Sub-alpine heaths and bogs p. 41

Mountain beech forest p. 38

Rivers and river gorges p. 45

Causses and karst p. 49

Rivers and river gorges p. 45

For the sake of clarity and simplicity – you are on holiday after all – we have reduced the 55 habitat types to a more digestible number of five general habitat types that can easily be distinguished and in which anyone wanting to explore the Cévennes and Grands Causses should spend some time. These habitats are scrubland (described on page 30), forests (page 34), subalpine heathlands (page 41) rivers and river gorges (page 45) and limestone plateaux (page 49). In most mountain ranges there is a clear distinction in habitat as one proceeds from valley to summit. But in the Cévennes, the same habitat seems to reappear – in different varieties– at different altitudes or on different soil types. Scrubland, for example, is a typical landscape type of lowlands with a distinct Mediterranean climate, but it reappears higher up on the mountains in a cold-and-wet adapted variant dominated by Piorno Broom* *(Cytisus oromediterraneus)*. Similarly, heathland is typical of the high, wet slopes of the Mont Lozère, but a version adapted to dry and hot conditions, dominated by Heather and Bell Heather, occurs further down the slopes. Similarly, a whole array of woodland types occur between valley and summit.

In the detailed descriptions further on, we will pay attention to the variety within the general habitats and mention some of the key species that occur there. For a more thorough description of the species in the Cévennes, read the flora and fauna section on page 71.

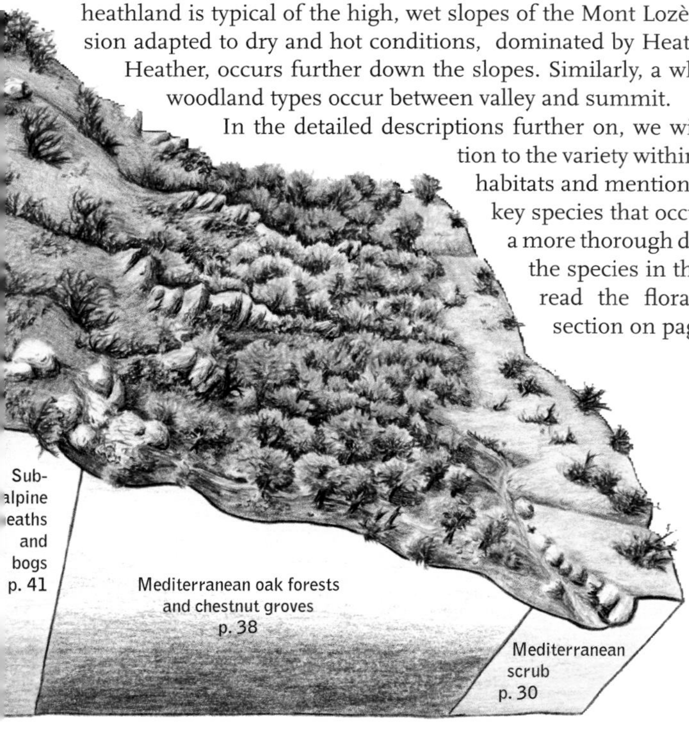

Sub-
alpine
heaths
and
bogs
p. 41

Mediterranean oak forests
and chestnut groves
p. 38

Mediterranean
scrub
p. 30

Mediterranean scrub

Mediterranean scrublands of limestone soils are found patchily on most Causse routes, but sizable tracts are found along routes 11 and 19. Scrubland of acidic soils are found on route 17. Broom scrub on high altitudes are found along routes 1, 3 and 9. The mountain heathlands are primarily found on route 3.

The south-facing slopes and the hills at the southern and eastern edges of the Cévennes are clad in a scrubland typical of the Mediterranean region. Here the climate is warm and dry, and plants of the higher regions can't survive the periodic drought and heat.

Hot and dry scrubland in the Gorge de la Vis (route 19).

In the eyes of most people, Mediterranean scrubland is a sort of uninteresting wasteland. In summer, it looks uninviting, dry, hot, dusty and thorny, the kind of place where a walk will soon become a test of endurance. In former days, scrublands were used only for marginal grazing, to collect brushwood and for beekeeping. Today, these areas are mostly abandoned, but an appreciation of their natural value and beauty is slowly growing.

For the naturalist, Mediterranean scrublands are superb places, primarily because they are home to a great variety of plants, butterflies, birds and reptiles. Many of them belong to species groups that are unfamiliar to the visitor from the north, such as the abundant flowery shrubs belonging to the family of the rockroses, the spectacular Two-tailed Pasha Butterfly, a European relative of a butterfly family of tropical Africa, and the big Cicadas with their deafening chirps. Those species that do have relatives in more northern regions, are quite often distinctly different. The Tree Heath, for example, is a lookalike of Heather, but can reach a height of six metres. The large Mediterranean Spurge and some of the umbellifers are in a different class of robustness than their relatives in North-west Europe.

Dartford Warblers frequent the dry evergreen scrublands on the edges of the Cévennes and Grands Causses.

Mediterranean scrublands are communities of survivors. Rather than competing with each other over scarce resources – as in most woodlands – the main struggle facing the inhabitants of scrubland is with the elements. Most species have strong defences against drought, the sun and herbivores (goats, sheep etc), which give them a tough and hardy appearance. Many of them have small leaves that are leathery, oily or hairy to prevent evaporation of precious water. Others, like Narrow-leaved Rue and Helichrysum or Curry Plant (so named because of the smell), have a strong taste and scent. This makes them less attractive for the goats and sheep.

Schist and limestone scrub

From a botanical perspective, the scrub of the acidic soil (mostly in the eastern Cévennes foothills; route 17) differs greatly from that of the limestone soils (both in the southwest and east; routes 11 and 19). Acidic soil is better in retaining moisture and thus has a more lush – as scrublands go – vegetation than its calcareous counterpart. Large shrubs of Tree Heath, Strawberry Tree (a small Mediterranean tree that is a relative of the heath family) Holm Oak and various cistuses occur here, like Sage-leaved and

Poplar-leaved Cistus grows in acidic scrublands.

Poplar-leaved Cistuses. Together, they form a dense thicket of vegetation with relatively few herbs.

Limestone scrub is usually less dense with a different set of species, such as Mediterranean Buckthorn, Etruscan Honeysuckle, Phoenician Juniper and many others. Amongst the shrubs there is usually an abundance of wildflowers (see page 80 for more details). In contrast to the flora, the fauna of limestone scrublands is rather similar to that of the acidic scrub. Only a handful of butterfly species, those that are bound to specific food plants, occur in one or the other types of scrubland. The birdlife is a mixture of typically Mediterranean species like Melodious and Subalpine Warblers and species familiar to temperate Europe, such as Blackcap, Blackbird and Red-backed Shrike.

In scrublands reptiles abound. The high temperatures, abundance of prey and ample hiding places, make them ideal for lizards and snakes. These are the places to search for Spanish Wall Lizards, Southern Smooth Snake and Montpellier Snakes, amongst others.

Other scrublands

Heather (small flowers) and Bell Heather (large flowers) colour the mountain heathlands in summer. Bell Heather flowers in July, Heather in August.

In the eastern Cévennes, the Mediterranean scrubland gives way, as one proceeds up the mountain, to true heather dominated scrub. Cévennes heathlands are dominated by two species of heath: Heather – the familiar species throughout north-west Europe, and Bell Heather, a truly Atlantic plant, that only occurs in Western Europe.

Unlike the Mediterranean scrublands, the heathlands flower in summer and the July and August visitor is greeted from afar by a sea of pink. Bell Heather starts to

Scrub of Piorno Broom* *(Cytisus oromediterraneus)* is found above 800 metres and visible from afar in May and June, when the shrubs flower.

flower in June and when it has peaked in July, Heather takes over. This late flowering period is, like in the heathlands in northwest Europe, of great importance to nectar collecting insects, such as bees. Consequently, the Honey Buzzard, who feeds on bees, is a typical bird of these heathlands and the surrounding woodlands.

At 800 metres and above, the heathlands become increasingly dominated by a species of Broom, the Piorno Broom* *(Cytisus oromediterraneus)*, slightly smaller and with deeper yellow flowers than the Common Broom. Piorno Broom is a typical species of the Spanish and South-French mountains, and some of the higher slopes of the Cévennes are completely dominated by this species. In other areas it is accompanied by Bracken Fern, and always by a thriving population of Greater Broomrape; a parasitic plant that taps into the roots of the Broom.

Like heathlands and Mediterranean scrublands, fields of broom establish themselves after a disturbance of some sort, often fire. Areas of broom persist as a stable vegetation type for a long time before giving way to forest. Uninterrupted broom fields harbour few birds, plants and insects, but they provide a sea of yellow blossom when the bushes burst into flower in late May and June. They are not just a feast for the eye but also for the nose, as broom has a wonderful scent.

Forests

Routes 2, 11 and 18 partially run through Downy Oak woodlands. Beautiful Chestnut forests are encountered along routes 3 and 13. Mountain forests of Beech and Spruce are particularly scenic along route 3 and on the northern and eastern slopes of the Lozère (route 7 among others). Pine forests are found along route 16.

The Cévennes abounds in forest. A glance at a map of the Cévennes shows that over half of the region is covered by woodlands of some sort, ranging from young, scrub-like trees to pine plantations and mature, broad-leaved forests.

This wasn't always the case. The majority of the green on the map refers to young woods that weren't there fifty or a hundred years ago. In the last decades of the 19th century (see history section on page 56) most forest was cut to fuel industries, warm houses and create agriculture land. With the abandonment of the marginal agriculture in the Cévennes, a policy of subsidized reforestation persuaded many farmers to plant forest on their land. There was spontaneous re-growth of forest as well, particularly on abandoned terraces and hillsides.

Today, one encounters forests of some sort from the Mediterranean lowlands and river valleys almost up to the summits of the Lozère and Aigoual. Only the highest parts are devoid of trees, but even here exceptions can be found in the small woodlands of stunted Beeches and Spruce plantations that grow on the Lozère.

The woodlands in the Cévennes are well worth exploring. Even the young forests, still half scrublands, support a variety of plants and animals. However, forest is here, as elsewhere, a habitat that hides its treasures well. Unlike the Causse grasslands, the interesting forest birds, plants, reptiles, insects and mammals that occur are local, rare or difficult to find.

The forests in the Cévennes vary in character depending on the soil and altitude. Four general types of woodland can be distinguished. First, there is the Mediterranean woodland in the valleys, with open and often young Downy Oak forests on the limestone soils. Its counterpart, the schist region, is dominated by Chestnuts and Holm Oaks. The second forest type occurs higher up: the extensive forests (often former plantations) of Sweet Chestnut. These are, almost without exception, found on schist soils. Still higher, the third type is the mountain forest present on the higher slopes

of the central mountains of Montagnes des Bougès, Mont Lozère and particularly around Mont Aigoual. Beech, Spruce and locally Sessile Oak are
predominant in these forests. Finally, the fourth forest type consists of
pine woodlands with a superb flora and fauna in the gorges and locally
on the Causses. In addition to these four natural or semi-natural forests,
there are plenty of regimented coniferous plantations which are much less
appealing.

Montane Beech-
woods on the Mont
Aigoual (route 5).

Mediterranean woodlands

Mediterranean woodlands come in two forms; the Downy Oak woodlands
in the warm valleys in the limestone regions and the dense Holm Oak
woods on the schists.

Downy Oak forests are mostly young forests originating from spontaneous
re-growth after abandonment of terraces. Older, more varied types occur
on rocky slopes where agriculture was never possible. Downy Oaks form
the lower belt of woodland in the gorges, giving way to Scots Pine and
occasionally Beech woods higher up on the cliffs. Downy Oak forests are
sub-Mediterranean forests, occurring widely in the mountains in the Mediterranean basin and in the lower regions on the edge of the Mediterranean

The Bois de Païolive (route 11) is a beautiful example of a sub-mediterranean Downy Oak forest (top). Southern Saw-tailed Bush-cricket (below) is a common Grass-hopper species in the forest.

region, like here in the Cévennes. Large areas of this forest type are found in the south-western sector, around the Durzon, Dourbie and Jonte Rivers. In the east, the Bois de Païolive is a beautiful example of a Downy Oak forest (route 11).

The Downy Oak looks like a gnarled and smaller leaved version of 'our' Pedunculate Oak. The leaf litter breaks down easily and allows a rich undergrowth to appear. On rocky soils and in other places where the oaks are well spaced, many of the more robust grassland plants occur. There are plenty of orchids around, particularly the helleborines such as Müller's and Dark-red Helleborines and Violet Limodore (see also page 87). The open and fresh appearance of these forests make them a joy to walk through.

In contrast to the Downy Oak forests, the Holm Oak forests in the lower zones of the Cévennes have more the character of high-growing and dense scrubland than of a true forest and a strict separation with the scrubland on schist soils is impossible. The evergreen Holm Oak is the dominant species, but, as in the scrublands, there are many other trees too, such as Tree Heath, Strawberry Tree, Sweet Chestnut and Sessile Oak.

These forests are primarily found in the southern, south-eastern and eastern sector of the Cévennes, but penetrate the area along the Gardon valleys. Due to the abandonment of agricultural land, the forest is on the increase, but the most interesting, older patches are rare.

Chestnut groves

The forests and groves of Sweet Chestnut are the most striking feature of the Cévennes. The trees in Sweet Chestnut forests are often spontaneous resproutings of pollarded stumps in largely abandoned groves. They grow in the lower sections of the mountains, roughly between an altitude of 400 and 1000 metres on acidic (non-limestone) soils. This calcifuge (limestone-fleeing) character of the Chestnut is typical and the sudden occurrence or disappearance of this tree, when you drive through the countryside, is a very reliable indicator that you have moved into an area with another type of bedrock. Only once did we find chestnuts on limestone soil, in the Bois de Païolive.

In the lower parts, the Chestnut shares its domain with Holm Oaks and Maritime Pines (the latter being planted), and higher up with Sessile Oak and Beech. Throughout the Chestnut area there are varying areas of heathland consisting of Heather and Bell Heather.

Old Chestnut grove in the interior Cévennes.

The shade and the poor soils make the Chestnut forests one of the poorest areas botanically. The old and decaying trees, however, are attractive to birds of the older forests. Nuthatches, Green, Great Spotted and, to a lesser extent, Black Woodpecker are numerous in the forests. They share the holes with Edible Dormouse, Red Squirrels and Tawny Owls. Buzzards and Honey Buzzards breed in the old trees and hunt over the heathlands. Goshawks and Sparrowhawks are present in good numbers as well. Wild Boars do particularly well here, feeding on the fallen chestnuts. They are not easily seen, because hunting made them shy.

Mostly however, these forests are a cultural heritage as testified by massive old trees on certain locations in the true heartland of the Cévennes (see page 62).

Mountain forests

At an altitude between roughly 900 metres and 1500 metres, just below the bald heads of Mont Lozère and Mont Aigoual, extensive forests of Beech and Spruce cover the slopes. In places, they are mingled with Fir, Larch and Sessile Oak. Again, most forest is fairly young and intensively used for timber production and to be quite frank, rather boring. But there are some interesting patches of mature forests as well, with plenty of dead wood on the forest floor and moss lying thickly over the rocks.

Spruce-Beech forest is a typical montane forest, shaped by cold, often snowy, winters, cool summers and sufficient precipitation from the Atlantic. This kind of forest is present in most European mountains, but in the half-Mediterranean Cévennes, they add a distinct northern feel.

Dead wood is essential to the forest ecosystem, which is why old forests are so much more valuable than young stands. On damp north slopes, where abundant moisture aids the breakdown process of wood, the most interesting forests are found. Decay is the often forgotten second half of the forest ecosystem. We tend to think in terms of what grows and flowers, but in forests, many living creatures are involved in some stage of the decaying process. The Stag Beetle is one familiar species that is associated with dead wood. Also some rare plants, including orchids like Coralroot and Ghost Orchid, occur on decaying wood, although they are rare in the Cévennes.

The Black Woodpecker nests in old forests. This really is the carpenter of the forest, the only species that is capable of making nest holes in live wood. It fulfils an important function as such, because it enables other tree hole breeders to follow. On the Aigoual, the Tengmalm's Owl, a northern species on the edge of its range, is the most noteworthy bird.

Springs and creeks in the forest are home to the beautiful black and yellow Fire Salamander, which is best encountered on moist spring days, when it is on expedition in the forest. Small, peaty clearings have interesting dragonflies, butterflies (very typical in these habitats is the Scarce Copper) and wildflowers.

Black Woodpeckers (opposite page) occur in old forests and are capable of drilling a hole in living wood. As such they perform an important role in providing housing for cavity-breeding animals.

Maritime Pines were planted on a large scale in the beginning of the 20th century to prevent soil erosion. They grow mostly at low altitudes in the eastern Cévennes.

Pine and spruce woods

Pine woods are plenty and made up of three pine species: Scots, Austrian and Maritime Pines. All 3 species are (probably) native, but most pine woods are far from natural, being planted in the second half of the 19th century (see page 65).

Pine Forests have been planted throughout the region, but plantations are particularly in evidence in some parts of the Causses (primarily Austrian Pines here), the eastern foothills of the Cévennes (Maritime Pine), the Bougès (Scots Pine) and, with 40% the highest amount of planted forest, on the Mont Aigoual. Here, on acidic soils at higher altitude, Spruce, Larch, Fir and some other conifer species are more common.

Ecologically, they are an impoverished version of the natural woodlands. However, the plantations are not totally devoid of interest to the visitor, since there are always still some of the species and natural features that make the natural forests so interesting. (see also page 87).

Natural pine stands are common in the gorges, like here in the Gorge du Tarn. Orchids abound here, with Common Spotted Orchid (left) and Violet Limodore (right) as two of the more frequent species.

Subalpine meadows, heath and bogs

> The subalpine zone is only reached on the Mont Lozère, routes 7, 8 and 9 and the Mont Aiguoul, route 5. The Mont Lozère is larger and has a much more diverse range of mountain habitats than does the Aigoual.

Walking through the hot, Mediterranean scrub on the edge of the Cévennes, it is hard to conceive that a little further up, on the Mont Aigoual and Mont Lozère, little cold streamlets cut through fresh meadows, surrounded by cool and mossy beech woods.

Large, rounded boulders and granite sediment characterise the rounded hills of the Lozère. This is a cold bedrock with very little nutrients, which is reflected in the vegetation. The hillsides are covered in broom thickets and meadows, while peat has accumulated over the years in the little depressions, which soaks up the rainwater like a sponge and releases it slowly in countless little streamlets.

A Large Red Damselfly got tangled up in the leaves of the insectivorous Round-leaved Sundew.

Together with Mont Aigoual, Mont Lozère forms the highest, coldest and most 'Alpine' part of the Cévennes. These upper parts belong to the subalpine zone, with several plants and animals that are typical of mountainous areas. Don't expect to find an Alpine landscape on the Lozère, though. The maximum altitude of only 1,699 metres and the rounded gentle slopes of the upper parts are nothing like the jagged heights of the Alps.

The most prominent feature of the Lozère and Aigoual is the fact that their summits rise above the tree line. On the southern slopes the tree line almost reaches up to the ridge, but on the northern, colder slope, the treeless area is much larger. Fierce winds and high precipitation (with about 1600 mm annually, the Lozère ranks among the wetter places of France) are typical on both mountains. Together with the cold, impenetrable granite core this creates a distinctive northern feel on these high slopes. Particularly in summer, the gurgling streamlets, raised bogs and flowery meadows that form the backdrop of any walk on the Lozère couldn't be more at odds with the hot and dry lowlands.

Mountain heathland and bogs on the Mont Lozère (route 7).

At less than 1700 metres, the treeline is not a climatic boundary and the so-called subalpine meadows above the treeline are not naturally established. Instead the Lozère and Aigoual were once cleared to accommodate the grazing sheep of the transhumance. Each spring, when the Mediterranean grasslands were drying out, the sheep and cattle was moved up the mountains. Using 'drailles', broad drover tracks that cut through the Cévennes, the livestock was moved up the mountains where the pastures had just shaken off the winter snow and were fresh and flowery. To accommodate the livestock the scanty woodlands on the summits were cleared and transformed into meadows. Close to the natural treeline, the vegetation resembles the natural succession nonetheless, and many plants, butterflies, dragonflies and birds found on the Mont Lozère are also typical of the 'true' Alpine meadows.

Patchwork of landscapes

The Lozère landscape is a mosaic of damp meadows, broom and boulder fields, bogs and mires, but the main component is heathland. These heathlands are quite different from the coastal heathlands along the Atlantic. Those familiar with either the French, British, Dutch or German lowland heaths will most probably notice the modest role Heather plays on the high slopes (unlike the heaths further down in the Cévennes, which are dominated by Heather and Bell Heather), and wonder about all those Bilberries. Heather and Bilberry are members of the same Heath family as are the other major components of the Lozère heath, Bogberry and Cowberry, justifying the name heathland. The relationship with the lowland heaths is also strengthened by the occurrence of species as Petty Whin, Heath Rush and Purple Moor-grass.

These heathlands have elements of two worlds, because some Alpine plants also put their stamp on the landscape, like the conspicuous yellow flowerheads of Austrian Leopard's-bane, the pretty blue Perennial Sheep's-bit, Alpine Clover and – in abundance – Alpine Lady's-mantle. A more Mediterranean touch on the Lozère is present in the form of extensive thickets of Piorno Broom* *(Cytisus oromediterraneus)*, which produce an almost obtrusive yellow in May.

These are the more dry habitats. Wetlands are also abundant on the Lozère, because – unlike the limestone Causses– the granite core of the Lozère doesn't permit water to penetrate, while the mildly rolling topography makes a quick run-off impossible. Much of the rainwater remains on the mountain in a complex system of bogs, mires, springs, meadows and streams.

The bogs are easily recognisable as flat areas with white tufts of Common and Hare's-tail Cottongrass. These bogs are very important for the whole of the Cévennes. The Bog Mosses (Sphagnum) are nature's sponge, holding up to ten times its weight in water. During the summer, long after the snow has melted, the bogs continue to release water, thereby feeding the rivers all through the summer (to the delight of the canoeists of the Tarn, who would otherwise be pushing pebbles all the way up to their pickup point).

The Moorland Hawker hunts in the afternoon. It typically flies at a height of about three metres along woodland margins or above heaths.

As in northern Europe, these bogs are good places for plants, but given their location this far south, the Lozère is of particular importance. Many species have their southernmost location here (and in the Pyrenees). Depending on the exact bog type you may encounter Heath Spotted and Lesser Butterfly Orchids, Beak sedges, Marsh Cinquefoil, Round-leaved Sundew and, although it is very rare, the beautiful Bog Asphodel.

Like the flora, the fauna of the Lozère represents both an Atlantic and an Alpine world. Birds like Water Pipits and Ring Ouzels, reptiles like Adder and Viviparous Lizards and Dragonflies like Large Red Damselfly and Moorland Hawker are all typical of colder regions. As can be expected, the butterfly fauna is attractive, with the typical mountain groups, like Ringlets and Fritillaries, well represented.

The subalpine habitat occurs on the Lozère as well as on the Aigoual, but the surface area on the Lozère is much larger and the diversity is much greater. The Aigoual has a nice patch of heathland, but lacks the bogs and mires and the granite boulder landscape that makes the Lozère so interesting.

Atlantic mountain heathlands

The Mont Lozère and Aigoual are not the only mountain systems where such a summit ecosystem has developed. Other mountain tops in the Massif Central, but also in the western Pyrenees, the Cantabrian Mountains in Spain, the Serra de Estrela in Portugal and a few others in this area, have similar vegetation and accompanying wildlife. These mountains are invariably made of acidic, nutrient-poor bedrock and enjoy a high precipitation which creates a vegetation that is different from other mountains and is therefore unique. Together they are grouped under the term *Atlantic Mountain Heathlands*. They are much richer in species than the lowland Atlantic heathlands, and reach their highest diversity in northernwestern Spain and Portugal. The Aigoual and Lozère form the south easternmost of these mountain heaths.

The Water Pipit breeds on the Mont Lozère.

Rivers and river gorges

> The most impressive limestone river gorges are seen along routes 2, 3, 11, 16 and 19. Rivers over acidic schist soil are part of route 1 and 17.

The rivers and the gorges through which they flow exhibit two completely different habitats. But since it is the river's eroding force that has created the gorge, the river is consequently trapped at the bottom. River and gorge are intrinsically linked to each other.

River gorges are without doubt the most scenic parts of the Cévennes. Particularly the ones in the limestone areas –the gorges of the Tarn, Jonte, Dourbie, Vis and Chassezac- are spectacular, because the river has created steep, vertical cliffs. These gorges are several hundred metres deep. The Gardons – the river valleys in the schist bedrock of the eastern Cévennes – are more V-shaped and not nearly as deep.

River gorges are unique ecosystems and are exceptionally rich in plants and animals. They would have been by far the most interesting places to visit, if they only weren't so inaccessible. But then again, if they weren't, they would probably have been cultivated and no longer as rich in plants and animals.

Certainly, one element that makes them so unique is that there are large areas that are completely wild, uncultivated and never visited. In the Gorge du Tarn for example, there are cliff forests with the character of primeval forest.

Cliffs of the Gorge de la Jonte (route 14).

Only in these almost completely inaccessible areas the superb Lady's Slipper grows, an unattainable lady indeed. But there is more to these gorges than mere inaccessibility. The sheltered location and the dampness from the constant presence of the river, protects the valley from extremes of heat or cold. The abundance of water and of nutrients brought down by the river allows a luxuriant vegetation of Alders, Ashes (both Common Ash and the south-European Narrow-leaved Ash) and large-leaved herbs. This is the proverbial front row: a favoured location for growth, but there are only limited 'seats' and consequently a constant struggle for a place.

When walking one of the gorge trails you will experience this lush vegetation, but sooner or later you will find yourself on a vantage point on a rocky protrusion. There on that windswept and bare spot, with the sun (or with less luck, the rain) beating down relentlessly, you quickly come to realise that the gorge as a sheltered location isn't the complete ecological story. More characteristic of the gorge than shelter is the diversity of conditions, from moist, nutrient-rich and sheltered sites to dry and exposed spots and everything in between (including less obvious combinations such as exposed, moist sites and sheltered dry places). River gorges are without doubt the most diverse areas in the Cévennes, and harbour patches of scrub, steppe vegetation, marshy vegetation and woodland within them. This diversity is particularly evident when looking at the shrubs and trees.

Gorge du Tarn

Almost every species of the Cévennes is found in the gorges, and often they are growing cheek by jowl of one other. Some species are confined to the gorges, such as Linden and various Maples.

Cliffs

As most dramatic and most inaccessible of all the habitats, cliffs
are particularly appealing to visitors. Attractive but frustrating too,
since you can never really explore
them. More than once we discovered a rare plant or butterfly hope-
lessly far out on a ledge, without
even the remotest chance of getting closer for a good look.

Two cliff dwell-
ers: Peregrine
Falcon (top) and
Sticky Columbine*
(*Aquilegia viscosa,*
bottom). The latter
is endemic to the
region.

It is precisely the inaccessibility to the more clumsy land-dwelling animals that decides what life is present on cliffs. Apart from a few very good climb-ers, only airborne animals and plants (through their seeds) can reach the cliffs. Here animals and birds are free from predators such as Stoats and Foxes. These conditions, plus the reliable updrafts of air, make cliffs impor-tant breeding sites for birds and bats. The Griffon Vulture (see text box on page 104) is without doubt the most visible of the cliff breeders, but Egyptian Vulture, Golden Eagle, Eagle Owl, Peregrine Falcon, Red-billed Chough, Blue Rock Thrush, Crag Martin and Alpine Swift are also typical cliff birds. The flora of cliffs is remarkably rich as well, with several species that are endemic to the Cévennes (see page 76).

River and river borders

The rivers of the Cévennes are beautiful and clean and on a warm day the banks are as inviting for dozing on as they are for exploring their flora and fauna. Except for the Tarn and some rivers in the extreme eastern part of the Cévennes, there is no canoeing, which makes the rivers quiet and unspoiled.

Streams and brooks are numerous in the region, but only outside the limits of this guidebook do they become more sizeable. This limits the number of particular birds within our area somewhat. Although typical river-dwelling species like Grey Wagtail and Dipper are present in good numbers, those of broader rivers like Kingfisher, Little Egret and Common Sandpiper are rare within the Cévennes (although quite numerous in those same rivers a little outside the mountains).

The Cévennes rivers invariably run through fairly narrow valleys and sometimes bordered by a narrow belt of gallery forest (so called because with the overhanging branches it appears like a gallery when seen from the river). Along the river, gravel banks are common and important ecological features as they break the stream, creating both rapids and shallow, calmer areas where the water can warm up. These latter areas are important for dragonflies, fish, and amphibians to lay eggs. It is also the place where young Viperine Snakes come to hunt for tadpoles. On moist edges of the gravel banks, butterflies (mostly Blues) come down to drink and dragonflies (mostly Pincentails and Damselflies) come down to rest, while Gravel Wolf Spiders* *(Arctosa cinerea)* hunt on the drier parts. Sometimes it is best to just sit and relax on such a gravel bank, and see what comes by. At specific places at dawn or dusk you may be rewarded by passing Beavers, which have been reintroduced and are now quite common (see page 99).

Gravel Wolf Spider* *(Arctosa cinerea)* has a cryptic pattern and is very difficult to spot unless it moves. Search for it on gravel banks in the river.

Causses – dry plateaux and karst

Routes 1, 2, 4 and 6 traverse Causse country by car. Routes 12, 14, 15 and 18 are hiking routes on the Causse.

The Causses are dry, limestone steppes with a unique flora and fauna.

Without doubt the Causses are the most striking and interesting of the Cévennes landscapes. The massive high-rise plateaux form a habitat that is unique in the whole of France.

The Causses were created in the same time as the Alps rose. The collision of the African and European tectonic plates pushed up a shallow sea bed to their present altitude of about 900 metres. This became the Causses; a thick limestone layer, originating from the former marine sediments.

The plateaux are characterised by rocky hills, covered by steppe grasslands, open scrub, alternated with coniferous forest. Water is absent, and there are no valleys cut out by streams. The depressions, called poljes or dolines (see page 24), have a richer and moister soil, where cereals are grown. As a habitat, the Causses are best compared with the steppes and steppe woodlands of Eastern Europe. The latitude would provide the region with a sub-Mediterranean climate but the altitude, ranging between 600 and 1,250 metres, ensures harsh winters, simulating a continental climate in a country that is primarily dominated by moist and mild weather from the Atlantic. The rainfall is much higher than that of the steppe regions of Eastern Europe, but in the limestone soil, the water sinks away, making the Causses a much drier habitat ecologically than precipitation graphs would suggest. All in all, it should come as no surprise that some of the plants and animals found on the Causses are much more typical of Eastern Europe.

Dolines are the only agricultural areas on the Causses. The fresh green area in the centre of the picture is an example.

Yet looking at the vegetation from up close, the visitor is more likely to be reminded of the plant paradises on the chalk hills of southern Britain, the limestone hills of southern Germany or the Swedish Island of Öland (depending on what your point of reference is). The flowery topsoil with its rockroses, its orchids, the scatter of Horse-shoe Vetch and Globularia plus a whole lot more, conjures up images of these places, only on a much larger scale and with even more species. Yet the asclaphids and the Praying Mantises and many of the butterflies on these grasslands show that the Mediterranean isn't far away.

All in all, the Causses are a unique mixture of species from several different corners of Europe.

These species do not occur at random of course. There are distinct landscape types to be found on the Causses, each with its own attractions. So far we've described the Causse as if treeless and certainly this 'Causse Nú' (nude Causse) is the most remarkable and attractive part. However, roughly a third of the plateaux is 'Causse Boissé'; in other words forested. The amount of woodlands differs from Causse to Causse, but they are always concentrated on the western part of

the plateaux. In contrast to the natural Downy Oak and Scots Pine woods, present in many of the gorges, most woodlands on the plateaux are Austrian Pine plantations and not very attractive for the visitor.

The open Causse is probably not naturally treeless, but summer drought, winter cold, high evaporation and generally rocky soils drives the climatological conditions close to the limit of where arboreal life is still possible. The high risk of forest fires (due to a combination of drought, wind and lightning) together with grazing pressure favours a dynamic steppe woodland. This means that if nothing happens, the open grasslands gradually become overgrown by shrubs and eventually trees, but this is a slow process and the disturbance of fire and grazing is sufficiently high to maintain a dynamic pattern of open grasslands, scrub and woodland. The

Corn Bunting (top) and Cirl Bunting (bottom) are two of the five species of bunting found on the Causses. The others are Rock Bunting, Ortolan Bunting and Yellowhammer.

high portion of grassland is attractive to herbivores, which is one possible explanation for the early settlement of hunter-gatherer tribes (see page 56). Later, the further clearance of the plateaux made the region perfect for tending sheep.

Beside grasses and herbs, shrubs do well on the Causses. Particularly, Gooseberry, Box and Wild Juneberry or Snowy Mespilus and Whitebeam are scattered over the open Causse.

Lavognes – Lifeblood of the Causses

Living on the Causses has never been simple, for one because of the lack of water. Up on the high plateaux there are neither rivers nor lakes and most rainfall, approximately one metre per year, vanishes into the porous limestone. This water emerges eventually in the rivers, which flow deep down in the valleys and gorges, but it does not help the people and livestock who live several hundreds of metres higher on the Causses.

Water supply has remained a problem until very recently. The Templar town of La Couvertoirade, for example, did not have mains water until 1975. Throughout the centuries, several methods of conserving water were adopted, with almost every house having a system of wooden or terra cotta guttering channelling rainwater into a cistern below the building.

For the sheep, more accessible water sources were necessary and the idea of the lavogne, a type of pond, was born. The first lavognes probably date back to the Neolithic period which, in this region, is between 5000 and 2000 B.C. Once man started to farm in these remote and arid areas the number of lavognes grew enormously.

The first lavognes were made by lining natural hollows with clay which enabled the water to collect without draining away through the limestone. A very good example of this 'proto' lavogne can be found next to the road leading to Causse Bégon, about 400 metres before the village. It was an easy way to collect water although the depth of such pools was never great and they were very difficult to keep clean. Later, a second design was developed which used a naturally occurring slope of solid rock with a retaining wall being built at the lower end, forming a basin. At St. André de Vézines, on the Causse Noir, is one of the most accessible of the 'rock' lavognes, situated at the junction of the D124 and D203. By far the most common design seen today is that of the paved lavogne; one lined with flat-sided rocks which were made deeper to hold more water and could be cleaned much more easily. An example can be seen on the south side of La Couvertoirade (route 5). Many of the original earth-based ponds were rebuilt in this fashion and some, due to leakage, have even been lined with concrete.

Once sheep farming started to increase in the mid 1800s there was a greater need for lavognes to cater for the increasing flocks. This led to much repairing and renovating of older ponds together with the building of new ones.

Lavognes were used only for the livestock, never as a source of water for the people. Even when drinking the rainwater stored in the cisterns, there was always a danger of disease, making it much safer to drink wine – perhaps this was one of the few advantages of this period!

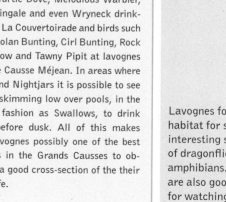

Lavognes do not only serve domesticated animals. They also attract much of the local wildlife and consequently are great observation sites. If you use your car as a hide or situate yourself behind a bush and wait, you should be rewarded with good sightings of birds, butterflies and dragonflies. We have seen Turtle Dove, Melodious Warbler, Nightingale and even Wryneck drinking at La Couvertoirade and birds such as Ortolan Bunting, Cirl Bunting, Rock Sparrow and Tawny Pipit at lavognes on the Causse Méjean. In areas where you find Nightjars it is possible to see them skimming low over pools, in the same fashion as Swallows, to drink just before dusk. All of this makes the lavognes possibly one of the best places in the Grands Causses to observe a good cross-section of the their wildlife.

Lavognes form the habitat for several interesting species of dragonflies and amphibians. They are also good sites for watching birds.

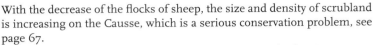

The White Rockrose is one of the most common wildflowers of the Causses.

With the decrease of the flocks of sheep, the size and density of scrubland is increasing on the Causse, which is a serious conservation problem, see page 67.

Where shrubs aren't colonising the land too aggressively, the Causse are a dream habitat for all naturalists. Cirl and Ortolan Buntings use the bushes as singing posts, Red-back Shrikes as hunting posts. The eerie call of the Stone Curlew echoes throughout the open vistas whilst a Montagu's Harrier pops in to hunt. Drifts of orchids and other botanical niceties are found in the thin grassland, visited by a large number of Fritillaries and Blues.

There are various types of open habitat, depending on soil depth, moisture, rockiness etc. The hilltops are often eroded, with rocky crests surfacing where the topsoil has eroded away. At certain locations, such as Nîmes le Vieux (route 15), Montpellier le Vieux and several places on the Causse Larzac and Causse Blandas (routes 4 and 6 respectively), this has created dry rock columns that rise from the open plain. These are great places for birdwatching, with specialties like Rock Thrush, Red-billed Chough and Rock Sparrow. Some of the endemic plant species are to be found on these sites as well (see page 76). On the dry soils, the typical grassland of the Causse is dominated by the beautiful Feather Grass, but interspersed with many other plant species. Orchids, in particular, do very well in these places with about 30 species that are either widely or locally abundant in the Causse (see page 94 and 220). On slightly deeper and moister soils, the thin grasslands become more meadow-like, with Upright Brome as the character-istic grass. More robust orchids, like Lizard, Pyramidal and Lady Orchid, abound here, together with an array of other plants.

Dark-red Pasque-flower* (*Pulsa-tilla rubra*) is one of several plants that are unique to the Causses.

The flora of the Causse with its abundance in species and varied structure of the vegetation, attracts an equally rich insect life. A plethora of butter-flies, moths, bees, bumblebees, grasshoppers, bugs, beetles and rose cha-fers are drawn to the flowers. To give an idea, in a typical meadow on the Causses it is possible to see around 200 specimens of up to 25 different species of butterflies on a good day. Predatory insects such as Ascalaphids (photo on cover) and Praying Mantis patrol the flowers for unwary insects. A quite successful method is deployed by the little crab spider which, suit-ably coloured, simply hides in the larger flowers, waiting for a bee to pay a visit. Stepping up a notch in the food chain, Wall and Green Lizards go after the larger insects, but in turn are prey to the various snakes that hunt in the early morning and evening. During the day the snakes hide be-neath the plentiful rocks to avoid becoming prey themselves to one of the many Short-toed Eagles that patrol the grasslands. In short, the Causse' grasslands support a complex food chain which makes them such a special habitat for naturalists.

History

Remarkable for such a "natural" area, the human history of the Cévennes is one of the longest of any part of Europe. From the earliest human presence almost until today, the story of this region has been one of an intimate, albeit ambivalent, relationship between Man and Nature. Nature has been Man's provider, but hardly a generous one. The constant struggle and hardships of scratching a living is the main theme of the region's history. Yet this economic poverty has endowed it with folklore, traditions and customs; all fathered by the spectacular, but harsh, natural landscape.

Prehistory

The oldest known records are from the Causse and date back to the early Stone Age, some 100 - 40,000 years ago. The extent of habitation of these earliest settlers is unknown; people were hunter-gatherers and the only evidence from that time is in the form of spearheads, pieces of pottery and the bones of the prey, such as Woolly-haired Rhino, Wild Horses, Aurochs

Menhir in the Feather-grass. Menhirs form the oldest evidence of human habitation in the region.

and Hyenas. Although the exact motives of these hunter-gatherers must remain a matter of speculation, it is likely that they chose to settle in the Grands Causse region because of the favourable combination of the presence of caves for shelter and large herbivores to hunt on the more or less open plains.

During the Middle and Late Stone Age, as the human population increased, so did the evidence of their presence. The many dolmens and menhirs that can still be found in the area are a sign that the human population had grown to, for those days, a high density, in particular on the Causse Méjean. This is remarkable because today Méjean is, with only 1.4 people per square kilometre, the least populated area of France. There are more dolmens (about 70) and menhirs (about 80) on the Méjean than there are hamlets, and probably more than there are inhabited houses. In 2005, the Méjean population stood at no more than 450 people. The dolmens were graves, but the significance of the menhirs remains uncertain, although it is widely accepted that they had some kind of religious significance, and, given their position close to the dolmens, possibly played an important role in burial rituals. You'll see many menhirs on route 10.

From the late Stone Age onwards agriculture came into existence as shown by traces of early agriculture on the Causse. However, the marginal soils and lack of water caused the mountain areas to lag behind in their development compared to the more prosperous valleys. The nearby Rhone valley and the Mediterranean plain of France became cultural hotspots. In the regions surrounding the Cévennes and Grands Causses, cities arose such as Nîmes, Arles, Avignon, and Montpellier, but, in contrast, the mountains themselves remained very much a wilderness; a position the Cévennes have maintained for a very long time.

Early history

The Cévennes' early history has not left many marks on today's landscape. Throughout the ages, the mountains were disputed between the peoples of the Mediterranean plain and those of the Rhone valley.

Around 1000 B.C., the Cévennes were occupied by the Celts. There were four different tribes, each with its own particular area. Then the Romans took possession of the region, with Caesar himself crossing the Cévennes in 52 B.C. on his way to attack the Gallic Arverni tribe, which lived a little further north in the Auvergne (a name that is derived from 'Arveni'). Caesar succeeded and the Romans occupied the region until 460 A.D. when the Franks (a Germanic tribe that eventually gave rise to the name 'France') and Visigoths (a Germanic tribe of Arian Christians) disputed occupancy of the southern parts whilst the Huns from Central Asia occupied the rest. The Visigoths eventually ruled supreme, utilising the buildings, writings and assets left by the Romans, but installing their own Arian bishops. Further invasions occured throughout the 7th to 10th centuries by the Moors who came from their base at Nîmes, eventually occupying the whole of the Cévennes from 720 A.D. Their arrival was a bloody one as they are said to have left no man, child or even animal alive. This occupation continued until 750 A.D. when the Moors left, allowing the Franks to take possession again.

This heralded a period of relative stability. The Roman Catholic religion was actively promoted throughout France and the Church grew in importance and in presence in the Cévennes. During the 11th century many large religious orders were established in the Cévennes, many due to overpopulation of the abbeys in the Languedoc and Ardèche. They received land, livestock and buildings from local lords in exchange for work, produce and, presumably, as a spiritual asset. By setting up farms around the abbeys, the local patchwork of farming as we know it today was established. This stability continued until the great schism divided Western Christianity between Catholic and Protestant. This had a severe impact on the history of the Cévennes.

Terracing hillsides was a popular method to create agriculture land in the Cévennes and Causse region. Today, most terraces have been abandoned.

The Huguenots

Like so many mountainous regions in Europe, the Cévennes has a history of rebellion and of outlaws. The wild and impenetrable character of the mountains made them good hiding places for those people or groups that were at odds with the authorities. The Cévennes sheltered the Huguenots. This group of Protestants withdrew into the Cevenol valleys when their Catholic oppressors wanted them dead. Dangerously close to Avignon, a Catholic stronghold, the Huguenots created a centre of protestant belief

Troglodyte houses – man-made walls shielding the entrance of a shallow cave – can be found here and there in the gorges. The houses in the picture can be admired on Causse Larzac (route 4).

at St. Jean du Gard. This precarious situation was temporarily resolved in 1598 when, in the Edict of Nantes, religious equality for the Huguenots and Catholics was formalised. The Catholics peacefully coexisted with the Protestants... for a while. Less than a century later, in 1685, this edict was revoked and relentless persecution of the protestant Huguenots followed. Many fled France for South Africa and North America with others seeking asylum in Britain, Ireland, The Netherlands, Germany and Scandinavia. The ones that stayed continued to practice their faith in secret. They used miniature bibles, which could be concealed in a woman's hairbun. This clandestine practice was risky. When two Huguenots were imprisoned by Abbot Chayla at Pont-de-Montvert, a group of people lead by Roland Laporte (born in St. Jean du Gard) and Jean Cavalier tried to procure their release, but in the ensuing struggles the Abbot lost his life. This started the Camisard Rebellion.

Many bloody battles between the Catholic establishment and protestant rebels, known as Camisards, ensued. The Camisards used the remote and vegetated Cévennes hills as their base and were supported by the Cevenol people, many of whom were Huguenots. This period marked the Cévennes as a region, since it eventually culminated in a furious Catholic king deciding to rid the Cévennes of all its inhabitants, Camisard or not. In September 1703 the Royal troops destroyed 466 hamlets and villages, thereby, of course, alienating the people even further. With Catholic partisans brought in to bolster the royal troops, chaos reigned with much looting and massacring taking place. Eventually, in 1704, Roland Laporte was betrayed to the troops and was killed at Castelnau-les-Valence. Although isolated acts of violence continued until 1710, this was effectively the end of any hostilities and the remaining Camisard leaders surrendered and left France. The outcome which followed this short but violent war of religion was that a protestant community was tolerated in the Cévennes, thus restoring peace to the region once more.

The arrival of the templars

Whereas the Cévennes developed as a battleground for religious rebellion, thereby showing the deep divide within the Christian world, the neighbouring Causse region took a very different road. Ironically, it developed and prospered as the logistic node for the violent export of western Christianity – the Crusades.

In 1120, long before the battles between the Catholics and the Camisards in the Cévennes, Hugues de Payns, a knight from the Champagne region, decided to set up a militia, which would protect pilgrims along their routes to the newly captured Holy Land. The Templar Knights' aim was to fight, of course, but also to live a life according to religious rule by adhering to the three vows of obedience, poverty and chastity.

By 1150 the Templars had become a large and famous order and had established themselves on the Causse Larzac at strategic points. The aim was to protect the trade and pilgrimage routes from the Causses down to the Mediterranean plain, particularly to the ports at St. Gilles and Aigues Mortes from where pilgrims could sail to the Holy Land. The village of St. Eulalie de Cernon on the Causse Larzac (the departure point of route 19) was the first place to come under the control of the Templars and they soon built the Commanderie and rebuilt the existing church. Over the years, La Cavalerie, La Couvertoirade and Viala du-Pas-de-Jaux were taken over and further developed by the Templars. Only in La Couvertoirade was an original Templar foundation. These villages, now splendidly restored, are today a must-see attraction for everyone who enjoys history and beautiful old architecture. The coming of the Templars meant prosperity for the local people. The locals were mostly farmers and herdsmen. The dry grassland of the Causses is suited for the grazing of sheep and the production of ewe's milk which was used to make cheese – including the famous Roquefort, which is the most important export product, even today. Under the firm rule of the Templars, there was little or no disturbance from bandits and farmers could tend their sheep and make cheese in peace. Moreover, the large numbers of pilgrims passing through the area ensured healthy sales.

St.-Eulalie-de-Cernon.

The Templar era covers two prosperous centuries of Causse history. In 1307, subsequent to the fall of Jerusalem, the Templars gradually fell from grace. The order was accused of heresies such as denying Christ, practising witchcraft, and many other un-Christian acts. The truth is that they had grown too rich and powerful for the envious secular authorities, so these former heroes were arrested, forced to confess and, in some cases, burnt at the stake.

All the property, which had belonged to the Templars was passed over to the Hospitallers, who spent little or no time at these sites on the Causses. Consequently defensive walls were built to protect the inhabitants from raiders and pillagers who, previously, would have not dared to attack due to the presence of the Templars. Nevertheless, the extensive agriculture and sheep cheese production of the Causses continued, which sustained the shepherds of the Causses.

The rise and fall of the chestnut culture

How different were the developments in the Cévennes! The Cevenol soil isn't suited for tending sheep on a large scale, and not for much else for that matter either. Only the high grounds of the Aigoual and Lozère were suitable for sheep grazing, but only during the summer. Each year, the sheep moved from the Mediterranean winter grounds in and around the Camargue up to the summer pastures on the Lozère and back again in autumn. This seasonal movement of stock is called the transhumance.

Farming in the lower Cévennes has always been a marginal existence and at a time of growing population in the 10th and 11th centuries, the poverty in the region increased. In response to this trend, the monastic culture in the region instigated new terracing and the cutting of forests in order to plant, graft and maintain the one crop that does well in the region, the Sweet Chestnut.

For centuries, the Sweet Chestnut was the base of existence for the Cévennes people. They call it l'Arbre du Pain, the Bread Tree.

This period marked the real rise of a chestnut culture which had already existed, on a limited scale, since Roman times and formed the basis of a successful system for several centuries to come. Chestnut cultivation became the most significant land use and the nut itself the prime source of nourishment for the Cevenols, with sheep herding, the growing of cereals and grapes and bee keeping as additional sources of income. The Chestnut saved the local population from famine on more than one occasion. Chestnuts were cultivated throughout the schist and – to lesser extent – on the granite soils of the Cévennes. The cultivation created dozens of varieties in the Cévennes alone, differing in flowering and ripening time, drought resistance and nut size.

Chestnuts were dried in special buildings, known as 'clèdes' (see next page). They were built on the hillside with their back ends towards the slope, so that what is first storey at front of the clède is at ground level at the back. In this way the chestnuts could be loaded at the back where they were dried over a smouldering fire. From chestnuts, every imaginable food product was made: a (very dark) bread was baked from its flour, sweetened chestnuts were eaten with it, roasted chestnuts were the usual snack and chestnut soup was served, often several times a day. Just as important was the tree. The wood was (and still is) used for window and door frames, the dried branches to make fire to dry the nuts and to keep the Cevenols warm during the winter. The 'green' branches were used in the making of baskets, cutlery and other kitchen tools. The nuts, harvested by entire families from mid-October to December, were so plentiful that there were still enough left to trade for wheat farmed on calcareous soils like the Causses. 'L'arbre du pain', the bread tree, thus fed and housed the Cévennes people for centuries. It was the sole pillar on which an entire existence was built; a living that was, whilst harsh, enough to allow for a steady population increase to reach a population density of some 35 people per square kilometre by 1850.

Silk and the renaissance of the Cévennes

The winter of 1709 was exceptionally cold. The frost wreaked havoc on the Chestnuts and many groves were destroyed or badly damaged. This turned out to be a fortunate accident, because it made room for planting another tree, the Mulberry. This small tree is the sole food plant of the Silk Moth. It started a modest Golden Age of the Cévennes, in which the region grew to become a considerable nucleus of silk production. Although Mulberries were planted in southern France as early as the second half of the 16th century, the industrious production in the Cévennes was heralded by the destruction of chestnuts in that winter of 1709.

For a little more than a century the Cévennes Silk was famed and transported all over the western world. The silk put the Cévennes on the European map. At last a profitable crop that would grow in the region! Everywhere Mulberries were planted and little silk farms were set up, often at the expense of the old chestnuts. The southeastern part of the Cévennes, particularly around Anduze and St. Jean du Gard, silk brought wealth to the Cevenols.

The museum of the Cevenol Valleys in St. Jean du Gard gives an insightful overview of the life in the Cévennes in those days, including an exhibition of the equipment that was used to process the chestnuts and the silk.

Crisis

There is always a danger in putting all your eggs in one basket, but the misfortune that awaited the Cevenols was exceptionally swift. Within a few decades, both economic pillars that had made the region prosperous for a short while were wiped out by disease.

An epidemic affecting the Silk Moths spread in 1847 and destroyed the silk industry completely. Louis Pasteur, the founding father of bacteriology, took the fate of the Cevenols at heart and worked to overcome the epidemic. Pasteur succeeded eventually, but the blow to the silk industry had been so great that it couldn't recuperate. With the strong competition from the East, the price of the silk collapsed and the final death blow was inflicted by competition from artificial fabrics. The last silk farm closed in the Vallee Française in 1968.

A few years after the silk moth epidemic, in 1870, the ink disease – la maladie de l'encre – hit the Chestnuts of the Cévennes and heralded a truly ink black page in the history of the area. The disease, a fungus, provokes a blackish secretion from the tree which then dies from the top down –without hope of recuperation.

The clède is a shed for roasting chestnuts. It is built onto a hillside so that the chestnuts could be loaded through the back, directly onto the first floor. The charcoal fire on the ground floor roasts the chestnuts.

Farmers saw their income destroyed and tried to compensate by selling damaged nuts to the tanneries, by switching to grape cultivation or by increasing the size of their flocks of sheep – activities that could only be marginally supported by the Cevenol soil. At the end of the 19th and the early 20th century, the population of the Cévennes dropped rapidly. By 1911 the population had decreased to 20 per square kilometre.

About forty years later, another tree disease, cancer of the bark, again a fungus that kills the tree from the top, caused yet more problems, although this disease proved to be treatable. Nevertheless, more people moved away in the hope of a better living elsewhere. In the process, many hamlets were deserted, dying chestnut groves left behind and terraced slopes became overgrown with weeds, then bushes and finally forest. The result of this exodus is still visible today. Anyone walking in the Cévennes will eventually come across a deserted clède in the middle of a forest or find ancient terraces in locations far from villages or towns. Along the smaller tarmac roads, many of the hamlets are deserted and empty.

In answer to these problems, farmers were subsidised to plant trees on their land. On the one hand this was to prevent soil erosion on the now bare and sometimes overgrazed lands, and on the other hand timber was to become a new source of income. There are several arboretums in the region, some of which are open to the public, where experiments have been conducted to determine which would be the best trees to plant. In the end, Maritime and Scots pines were planted in large numbers in the lower Cévennes, Scots and Austrian Pine on the Causses and a variety of mainly coniferous trees higher up on the Aigoual and Lozère.

Old, half overgrown trails often lead to deserted houses and clèdes. They are remnants of the time when the population of the Cévennes was larger than today.

66

The modern era

However, the tide has now turned for the Cévennes. Our modern age has brought high mobility and increased living standards in the whole of France and beyond. The Cévennes, it turns out, possesses an important product earlier generations knew, but never guessed would be worth anything; stunning scenery and a splendid natural wealth. Tourism has developed in this beautiful region and, with tourism, came money and the ability to import products from elsewhere. Prosperity came back to the Cévennes and, at last, the population is growing again albeit slowly. Our modern age has reversed the role of Nature and Man. No longer is Nature the stern un-yielding mistress of the Cevenol people, but has become the playful nanny that entertains the herds of hiking, biking and kite-surfing tourists. But now that the need for chestnut gathering and long days of transhumance through the cold and dangerous mountains are no longer needed, the Cev-enols are looking for ways to save their traditions and reaffirm their long

The Gorge du Tarn has become a very popular tourist destination because of its spectacular scenery and suita-bility for canoeing.

and profound relationship with their natural surroundings.

Tourism proves to be a helpful tool in this, reviving the traditional chest-nut- based industries. The products – liquor, icecream, marmelade, wood carvings, etc – are sold to visitors.

And what better way than to celebrate the end of a great day with a Chest-nut-flavoured icecream (Glace Marron) or a glass or two of the delicious chestnut aperitif? Bon appetit!

Nature conservation

The core of the Cévennes, including a small part of the Causses, is protected in the Cévennes National Park (CNP) and has the status of UNESCO Bio-sphere Reserve. The CNP does not own all the land within its boundaries, but works very closely with the landowners. Any reintroduction scheme or species protection plan is discussed and implemented with them in the form of a contract, ensuring maximum co-operation. Similarly, the CNP controls hunting to a degree by implementing 'No Hunting' areas (13,500 hectares) and regulates the 'harvest' of game species.

Outside this central core, both the periphery and the Grands Causses (which is designated a Parc Régional) enjoy only limited protection. This is quite incredible when you consider that in this area there are over 100 threatened and 33 endangered species, including over 50 endemic wildflowers (including two endemic orchids).

Within the river systems a successful 'clean-up' operation has been implemented to stop the discharge of waste from the various wool mills (on the Causses), slaughter houses and tanneries. On the Grands Causses, a policy was adopted to promote traditional farming meth-ods which makes no use of herbicides and pesticides. Unfortunately, the farmers are increasingly turning their backs on this scheme in an effort to make ends meet. As a result, the better areas are increasingly 'improved' with fertilisers. This destroys the natural herb-rich vegeta-tion (including the orchids) and the fertilisers subsequently run off in the local streams, causing problems in low-lying areas. Agricultural land that is too poor to make a living is abandoned, and this isn't good for the flora and fauna either. Open areas become overgrown by the natural re-growth of shrubs and trees, thereby losing their valuable steppe-like character with its rare birds, plants and butterflies. The open areas are disappearing at a rate of 1% per year, which does not sound like much until you realise that it has been going on for the

New Chestnut plantations are introduced to revive the ailing old groves and save the Cévennes Chestnut culture for generations to come.

last 20 years (Hence it is important to buy the products of traditional, non-intensive farming methods when you are there)! Tourism plays a double role in the conservation of the region. Not all tourism in the Cévennes is ecotourism in the sense that it is nature friendly and, particularly, in the Gorge du Tarn, the canoeing and other noisy sportive activities take their toll. Fortunately, the reintroduction of the vultures into the Jonte and Tarn Gorges has meant that these areas have been designated a protected area under the European Birds Directive. This means that indiscriminate rock-climbing and paragliding can be stopped in this area if it threatens the birds.

However, there is no doubt that the great touristic appeal of the Cévennes has led to a re-evaluation of nature. All this attention from tourists shows that the region's natural environment has an intrinsic value of its own and, much more important for the locals, has a value that can be translated into economic gain. This positive role of tourism could be still strengthened, by linking tourism to the traditional rural economy. Rural gites (guest houses), buying of local products from traditional farmers, greatly contributes to the preservation of this region, and, of course, to the visitor's own experience of the Cévennes. Some hints (both gastronomically and in terms of conservation) are given on page 218. See there how your visit can support the conservation of the Cévennes in a more concrete way.

Przewalski Horses on Causse Méjean.

Przewalski Horses

Le Villaret, a small deserted hamlet on the barest part of the Causse Méjean (route 2), has become the centre of a very important breeding programme for the rarest horse in the world, the Przewalski's Horse. It is the world's last remaining race of horses, which is truly wild (except for the Wild Ass and Zebras). The Mongolian steppes are the natural home of Przewalski's Horse. The Causse Méjean was chosen for this scheme because of its remoteness and its similarity to the grasslands of Mongolia. The programme's aim is to let the Méjean herd grow naturally, and then reintroduce the young horses back into Mongolia, hence preserving an endangered breed.

The project started in 1983. Four stallions were taken to the Cévennes on a trial scheme to see if the horses could cope on the Causse. It failed completely. Three died from congenital problems, possibly due to inbreeding at the zoos, from which they had been procured. They just weren't fit enough to survive in a natural environment. The fourth suffered, quite literally, a stroke of bad luck; during a thunderstorm he was killed by lightning.

In 1989 another attempt was made, which was successful. Eleven horses were brought in from Marwell Zoo in England. The WWF, in conjunction with TAKH, the Mongolian organisation dedicated to maintaining the breed in the wild, set the horses free to roam the Méjean. The following year the horses' area was more than doubled and they began to develop their natural social structure. Offspring followed soon thereafter.

The first twelve young horses were flown to Mongolia in September 2004 and a further ten followed in August 2005. Within a few weeks they had adapted completely to their new surroundings and eventually became the founders of a new population in their natural home.

Their years spent at Le Villaret in harsh winter conditions had stood these horses in good stead and they had learnt to feed on plants by scratching through snow if

necessary. This ensured their chance of survival in Mongolia. As of January 2008 the largest family group has five yearlings and four foals running with them and there are other young horses in the smaller groups – nothing short of a great success.

The Le Villaret herd will continue to live on Causse Méjean, and the young horses, which are born each year, will be used to bolster future re-introduction schemes and also to help strengthen the gene pool of herds kept in zoos across the world. It is possible to view the horses from the roadside at Le Villaret where there is a large parking area (see route 2).

FLORA AND FAUNA

Situated at a crossroads of the Mediterranean, temperate European and Alpine realms, the Cévennes and Grands Causses boast a high diversity of plants and animals. In particular, the variety of wildflowers and insects is exceptional. The varied geology (soil types and 'rockiness' of the soil) and geography (exposure to sun, wind and the altitude differences) creates many different little habitats, in each of which, different species thrive.
The Cévennes is therefore an exciting destination for the botanist, lepi-dopterist, herpetologist, ornithologist, naturalist and any other 'ists' with a wish list of species to find. But also those with a more general approach to nature will find plenty of species to stop them in their tracks and trigger their curiosity so unleashing that human trait of needing to know what, exactly, it is (well, that goes for us at least).

In the Cévennes and Grands Causses five large eco-regions come together, each contributing its own flora and fauna to the Cévennes landscape. These are the Mediterranean, the temperate (central European), the Atlantic and the Alpine regions. The fifth is a local eco-region and, as such, the most exceptional because it is characterised by an assemblage of species that is unique in the world. We are of course talking of the limestone upland steppe, or Causse region, with its unique mixture of Mediterranean species, Eastern European steppe species and species that are completely unique to the Causses. The French vegetation classification (CORINE biotopes) mentions several types of habitat that are only found on the Causses. This is based on the vegetation alone but – in our opinion – the flora and fauna combined are sufficiently distinct to call it a fifth ecozone. The examples of endemic Causse species are numerous. Most of them are plants, the most emblematic being the two endemic orchid species, Aymonin's Orchid* and Aveyron Orchid*. But Cévennes Saxifrage* and Sticky Columbine* are equally good examples. There are endemic species away from the Causses as well, although not too many. Cévennes Thyme is a very common and eye-catching example in the heathlands of the eastern Cévennes.

A Blue-spot Hair-streak on a Curry Plant. The butterfly fauna and the flora are particularly rich in the Cévennes.

The bulk of the flora and fauna in the Cévennes consists of Mediterranean and temperate European species, in addition to the 'sub-Mediterranean' flora and fauna, which is typical of the contact zone between the temperate and Mediterranean world.

Mediterranean species are, as one would expect, mostly present in the southern and eastern edges of the region and only spread into the core area along the lower valleys. The vegetation here consists almost completely of Mediterranean evergreen shrubs and trees (including various species of Rockroses, Brooms and Strawberry Tree) with wildflowers blooming mostly in spring. The birdlife too includes distinctly Mediterranean species like Woodchat Shrike and Short-toed Eagle, making the area of interest to birdwatchers. The temperate element is found higher up in the mountains, for example the typical assembly of anemones, Corydalis and Moschatel flowers in the Beech woods of the Mont Lozère.

Between the Mediterranean and temperate European regions, there is a distinct set of species that seems to belong to neither one nor the other, but thrive in intermediate conditions These 'sub-mediterranean' species are well represented in the Cévennes. Most orchids, for example, occur only in the cooler parts of the Mediterranean (e.g. Military, Monkey, Bee, etc.), while in temperate Europe they are restricted to the warmer regions. The Western Whip Snake is another example, as are two localised dragonflies, the Pronged Clubtail and the – indeed splendid – Splendid Chaser.

The upper reaches of the mountains are reserved for the Atlantic and the Alpine elements of Cévennes' flora and fauna: the broom fields and heathlands that are typical of that edge of Europe which borders the Atlantic Ocean. One typical, and very common representative of these Atlantic species, is the Bell Heather that dominates the heathlands in the eastern Cévennes.

Interestingly, the Alpine species found in the Cévennes occur at much lower altitudes than they do in the Alps. At 1600 metres, you can find Alpine Clovers and Spring Pasqueflowers for which you need to climb much higher in the Alps. The Lozère also has a number of Alpine birds, such as Citril Finch and Ring Ouzel. On the other hand, it should be mentioned that many species that are widespread in the Alps are not present in the Cévennes.

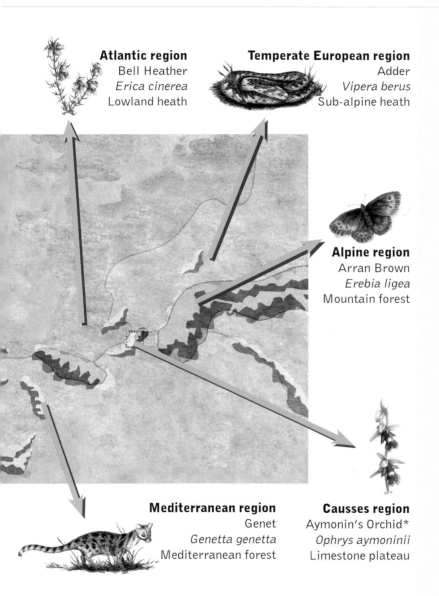

Atlantic region
Bell Heather
Erica cinerea
Lowland heath

Temperate European region
Adder
Vipera berus
Sub-alpine heath

Alpine region
Arran Brown
Erebia ligea
Mountain forest

Mediterranean region
Genet
Genetta genetta
Mediterranean forest

Causses region
Aymonin's Orchid*
Ophrys aymoninii
Limestone plateau

Flora

The most interesting routes from a botanical perspective are routes 2, 14, 16 and 18, all of which are on limestone soil. The sub-alpine flora is richest on routes 7 and 9. For the flora of the schist landscape, try route 3 or 17.

The flora of the Cévennes can be summarised in one word – superb. The Cevenol soil is gifted with the ability to bring forth the most beautiful and amazing species of wildflowers and bring them forth in profusion. The topsoil layer of the Causses, thin and rocky as it may be, produces a diversity of species that would make any botanical garden director envious. Without exaggeration one can say that the Cévennes and Grands Causses are one of Europe's botanical treasure troves, and one that lies only a day's drive from Britain, Germany or the Netherlands.

The first wildflowers appear in late March and the show continues all through the summer and well into autumn. The peak, however, is in May and June, when you encounter vast drifts of Wild and Pheasant's-eye Daffodils up on the cool mountains, rocky slopes turn into rock gardens, while

A flowery meadow on the Causse Larzac.

the Mediterranean scrub seduces you from afar with the herbal scent of Rosemary, Thyme and Marjoram. Most of the orchids – and there's nearly 60 species of them – flower in May and many of them in abundance (see page 94 for a separate section on orchids).

What is also great about the flora of this area is that it is not only appealing to the dedicated botanist, but also, since many flowers are so attractive, to anyone with an aesthetic sensibility. Finally we should mention the benefit of the presence of good identification guides, which are readily available (see page 217) – in other words, if you enjoy wildflowers, the Cévennes is the place to go.

So many species

Within the National Park boundaries there are some 2200 species of vascular plants. This is three quarters as large as Germany's flora, roughly the same as that of the UK and about one and a half times that of the Netherlands. Since many floristically superb sites are outside the National Park, the total number of species of the entire region covered in this guidebook might well exceed the 3,000 mark.

There are three main reasons for this fabulous richness. The first is the diversity in soil types that host different plants. Secondly, there are many different microclimates in a small area due to the variation in altitude and exposure to the elements. And thirdly, the position of the Cévennes as a contact zone of eco-regions (see page 73) ensures the presence of plants from the Mediterranean as well as from the temperate, Atlantic and Alpine plant regions. This contact zone is not restricted to the Cévennes. Rather it is a belt that runs from the southern and eastern Pyrenees past the eastern and southern rim of this region, through the Balkans to northern Greece. All mountain ranges in this belt exhibit an extremely high level of plant diversity with good numbers of endemic species. The eastern Pyrenees and Provence boast even more species than the Cévennes.

Some good habitats

Where to head to for the best plants? From the above introduction one could be mistaken in thinking that it doesn't really matter where you go in the Cévennes to find wildflowers. Although it is indeed hard to pinpoint a particular place that stands out above all others, some habitats are clearly richer than others.

The clearest division is between the limestone regions of the Grands Causses, the Schist Cévennes and the Granite cores of the Mont Aigoual and Mont Lozère. Botanically, these are like three different worlds and we advise anyone with an interest in wildflowers to spend time in all three. The limestone regions are the most interesting because they have the highest number of species, plus they abound in the more charismatic ones like orchids and lilies.

Since it is impossible to mention all species, we will discuss the flora of the botanically most interesting habitats with reference to the routes and the most attractive species that can be found there. A more comprehensive list of plants is given at the end of each section.

Flora of cliffs and rock slopes

One of the most superb habitats in which to indulge in some serious wild-flower spotting are the abundant cliffs and rock slopes of the region. The first places that come to mind are the impressive limestone rock walls of the Tarn, Jonte and Vis Rivers. However, equally superb and often more accessible are the karst pillars and smaller cliffs on the Causses.

Flora of limestone cliffs (routes 2, 15, 16, 19)

Many of the endemics, species which are confined (or nearly so), to the Cévennes, are found on limestone rock cliffs. Lecoque's Red Valerian* *(Centhranthus lecoqui)*, for example, has a limited range, from the Cévennes to Spain, but is quite a common species in the lower, hot and exposed parts of the limestone gorges. It often grows right on the roadside, together with some beautiful, more widespread species, such as Mountain Lettuce, Blue Aphyllanthes, Beautiful Flax, Snapdragon and Pale Stonecrop. All except Pale Stonecrop, flower in May and June.

Limestone cliffs on the Causse de Larzac. These sites are great for finding wildflowers.

Typical limestone cliff plants

Cluster-flowered Sandwort* (Arenaria aggregata), Lecoque's Red Valerian (Cethrantus lecoquii), Mountain Lettuce (Lactuca perennis), Blue Aphyllanthes (Aphyllanthes monspeliensis), Beautiful Flax (Linum narbonense), Snapdragon (Anthirrinum majus), Prostrate Toadflax (Linaria supina), Malling Toadflax (Chaenorrhinum origanifolium), Pale Stonecrop (Sedum sediforme), French Sermountain (Laserpitium gallicum), Common Sermountain (L. siler) Cretan Athamanta (Athamanta cretensis), Proliferous Pink (Petrorhagia prolifera), Wall Germander (Teucrium chamaedrys), Rouyan's Felty Germander* (T. rouyanum), Golden Felty Germander (T. aureum), Perennial Yellow-woundwort (Stachys recta), Yellow-wort (Blackstonia perfoliata), Leuzia (Leuzia conifera) Staehelina (Staehelina dubia), Blue Catananche (Catananche caerulea), Purple Viper's-grass (Scorzonera purpurea), Spanish Gorse (Genista hispanica), Silver-lined Broom* (Argyrolobium zanonii), Cévennes Saxifrage* (Saxifraga cebennis), Sticky Columbine (Aquilegia viscosa), Cévennes Cinquefoil* (Potentilla caulescens cebennis), Large-fruited Alyssum (Hormatophylla macrocarpa), Yellow Whitlow-grass (Draba aizoides), Kernera (Kernera saxatilis), Pungent Pink* (Dianthus pungens) Yellow Leek (Allium flavum), Rock Candytuft* (Iberis saxatilis), Prost's Candytuft* (Iberis prostii), Annual Candytuft (Iberis pinnata), Rock Soapwort (Saponaria ocymoides), Pyrenean Bellflower (Campanula speciosa)

At the sheltered bottom of the cliffs, spring is the main flowering time. One could spend days exploring the rocks, if only the sites weren't so difficult to reach! In particular the Pink family (e.g. Rock Soapwort, Proliferous Pink), the Mint family (e.g. Perennial Yellow-woundwort, Wall Germander), the umbellifers (e.g. French and Common Sermountains) and the daisy family (e.g. Blue Catananche, Staehelina) are well-represented on the cliffs.

Moving up towards the higher section of the cliff, the vegetation changes. The upper parts of the cliffs are exposed to remarkable environmental conditions: on one hand, the degree of solar radiation and the lack of precipitation reflect the Mediterranean conditions. On the

A typical species of shady limestone cliffs; Fairy Foxglove.

other hand, the temperature fluctuations and strong winds fit a more steppe-type climate. As a result, the vegetation consists of a mixture of temperate European, Alpine, Mediterranean and steppe species, plus a fair number, which have evolved here to adapt to this particular situation. Together with the karst rocks on the Causse, the upper region of the cliffs supports the highest numbers of endemic species. This is for example the haunt for the rare endemic Cévennes Saxifrage* *(Saxifraga cebennis)* and Sticky Columbine* *(Aquilegia viscosa;* p. 48), which is much smaller than the abundant Common Columbine. The small, white-flowering Cévennes Cinquefoil *(Potentilla caulescens cebennis)* is another endemic species. The Alpine Mezereon and Pyrenean Bellflower, a spectacular bellflower with a big stalk of large, bright-blue bells, are found on the upper cliffs.

Flora of schist cliffs (routes 1, 3)

The schist-rock flora is not as species rich as the limestone cliffs, but still there is a lot to discover. The thin blue spikes of Daisy-leaved Toadflax, the striped, pink flowers of the Pale Toadflax and the large, creamy flowers of the Creeping Snapdragon are all a delight to the eye. You'll find them to be numerous on rocky outcrops and on the edges of heathlands, often in the company of Crested Knapweed* *(Centaurea pectinata)* which has conspicuous outward curved flower bracts.

But above all, the schist rocks are the territory of the Stonecrop family *(Crassulacea)*. There are no less than 12 species of this otherwise none too diverse family in the schist Cévennes. Among them are the well-known Biting Stonecrop and White Stonecrop, but also lesser known species like the small Thick-leaved Stonecrop* *(Sedum desaphylla)*, which has swollen leaves that look like water drops, and the similar, if slightly larger, Short-leaved Stonecrop* *(S. brevifolium)*.

Schist cliffs support many species, like Rock Chamomile, Cévennes Thyme, Pale Toadflax and various stonecrops, like the Thick-leaved Stonecrop on the facing page.

The houseleeks – peculiar plants typical of the Alps – belong to the Stone-crop family as well. Two species occur, the large Mountain Houseleek and the smaller Cobweb Houseleek, which has "hairs" overlying their rosettes that look a like a spider-web.

Ferns too, grow from the cracks of the schist and from old walls. Apart from the widespread Rusty-back Fern, Maidenhair Spleenwort and Wall Rue, there are some rarer species , like Forez Spleen-wort* *(Asplenium foreziense)*.

Endemic species on the schist rocks of the Cévennes are few and two deserve special men-tion: the pretty Cévennes Thyme* *(Thymus nitens cebennis)* which is readily distinguished by its large and dense flower stalks of pale flowers and the Granite Pink* *(Dianthus graniticus)* which, as the name suggests, also grows on granite soils.

Thick-leaved Stone-crop.

Between the granite boulders on the Mont Lozère, one can find Forked and Maidenhair Spleenworts, Parsley Fern and, in shady places, Hard Fern and Common Polypody. At the higher altitudes, you may also find Prost's Saxi-frage* *(Saxifraga pedemontana prosti)*, Large-flowered Sandwort, Sesam-oides and Three-leaved Valerian.

Typical plants of schist cliffs and walls

Creeping Snapdragon (Asarina procumbens), Rock Chamomile* (Anthemis cre-tica), Rustyback Fern (Ceterach officinarum), Forez Spleenwort* (Asplenium foreziense), Maidenhair Spleenwort (A. trichomanes), Wall Rue (A. ruta-mu-raria), Jacquini's Mignonette* (Reseda jacquinii), Sesamoides (Sesamoides pygmaea), Livelong Saxifrage (Saxifraga paniculata), Granite Pink* (Dianthus graniticus), Proliferous Pink (Petrorhagia prolifera), Pale Toadflax (Linaria repens), Large-flowered Sandwort* (Arenaria montana), Daisy-leaved Toad-flax (Anarrhinum bellidifolium), Hoary Stonecrop* (Sedum hirsutum), White Stonecrop (S. album), Thick-leaved Stonecrop (S. dasyphyllum), Short-leaved Stonecrop (S. brevifolium), Pink Stonecrop (S. cepaea), Orpine (S. telephium), Navelwort (Umbilicus rupestris), Mountain Houseleek (Sempervivum monta-num), Cobweb Houseleek (S. ararchnoideum), Grass-leaved Plantain* (Plan-tago holosteum), Early Star-of-Bethlehem (Gagea bohemica), Clusius' Saxi-frage (Saxifraga clusii), Lamb's Succory (Arnoseris minima), Sticky Catchfly (Lychnis viscaria)

Flora of the Mediterranean scrubland (routes 6, 11, 17, 19)

Whereas in other vegetation zones the emphasis lies on the herbs, here the shrub are the main attraction. There is a large variety of small trees, shrubs and dwarf shrubs, many of which are aromatic, edible or have medicinal properties. Lavender, Thyme, Olive Tree, Mastic Tree, Rosemary and Juniper are all familiar from the garden, the kitchen and the medicine cabinet.

Many scrubland plants are thinly spread or only common in particular locations. Since the scrubland is a difficult terrain to work, and many species find shelter underneath the hardy shrubs, there is always the feeling that much more remains to be discovered. For example, several Mediterranean orchid species (see page 94) are listed for the scrublands on limestone soils, but they are on the whole quite rare and difficult to find.

As with all vegetation in the Cévennes, the scrublands on limestone soils support more different species than those on acidic schists. Conspicuous species on limestone are Kermes Oak, Mastic and Turpentine Trees, Mock Privet, Wild Olive, Mediterranean Buckthorn, Wild Jasmine, Smoke-tree, Scorpion Broom* *(Genista scorpius)*, Phoenician Juniper and Etruscan Honeysuckle (with beautiful pink-creamy flowers – p. 210). The smaller shrubs are easily outgrown by some of the larger herbs, such as Mediterranean Spurge and Narrow-leaved Rue* *(Ruta angustifolia* – a member of the buttercup family) and large umbellifers like French Sermountain and Cretan Athamanta.

On the ground, labiates and orchids are of particular interest, with Perennial Yellow-woundwort, Simplebeak Ironwort, Wild Thyme and Wild Marjoram in evidence. The most typical orchid of the scrubland is Woodcock

Mediterranean Scrubland in the Gorge de la Vis. This habitat supports a great range of species, including Orchids and Lilies.

Orchid, but other species of the Ophrys genus are frequent as well. Of the lily family, the robust White Asphodel and its smaller and more delicate cousin, the St. Bernard's Lily, are frequently encountered. In spring common Jonquil and Crimean Iris* *(Iris lutescens)* are present.

On schist (route 17), the Heath and the Rockrose families are particularly well represented. In the denser scrub, large heath species are Besom and Tree Heaths, whereas Heather and Bell Heather are the typical species of lower scrub. A typical small tree of this scrub is the Strawberry Tree, actually a member of the heath family, which is named after its edible fruit.
Rockroses (or Cistus) are frequent on schist soils. In spring, this typically Mediterranean family of plants boasts a large variety of flowery bushes, most of which are distinguishable by the shape of their leaves rather than their flowers. Of the white-flowering species, Sage-leaved, Poplar-leaved and Laurel-leaved Cistuses are the most numerous. Alison Scrub-Rockrose* *(Halimium alyssoides)* is the only yellow shrub-sized rockrose.

The large-flowered Spanish Broom is a conspicuous shrub in the Mediterranean plain.

Typical plants of limestone scrub

Phoenician Juniper (Juniperus phoenicea), Kermes oak (Quercus coccifera), Wild Peony (Paeonia officinalis, rare), Scorpion Broom* (Genista scorpius), Bladder Senna (Colutea arborescens), Etruscan Honeysuckle (Lonicera etrusca), Evergreen Honeysuckle (Lonicera implexa), Heath Fumana (Fumana ericoides), Simplebeak Ironwort (Sideritis romana), Perennial Yellow-woundwort (Stachys recta), Pale Stonecrop (Sedum sediforme), Orlaya (Orlaya grandiflora), Woodcock Orchis (Ophrys scolopax), Small Spider Orchid (Ophrys araneola), Yellow Bee Orchid (O. lutea), Pyramidal Orchid (Anacamptis pyramidalis), Lizard Orchid (Himantoglossum hircinum), Rush-leaved Jonquil (Narcissus assoanus), Crimean Iris (Iris lutescens), Wild Jasmine (Jasminum fruticans), Dwarf Scorpion-vetch (Coronilla minima), Common Lavender (Lavandula angustifolia) French Sermountain (Laserpitium gallicum), Cretan Athamanta (Athamanta cretensis), Mediterranean Spurge (Euphorbia characias), Nice Spurge (Euphorbia nicaeensis), Narrow-leaved Rue* (Ruta angustifolia), Butcher's Broom (Ruscus aculeatus), Common Smilax (Smilax aspera)

Typical plants of schist scrub
Hairy Adenocarpus* (Adenocarpus complicatus), Hairy Greenweed (Genista pilosa), Laurel-leaved Cistus (Cistus laurifolius), Poplar-leaved Cistus (Cistus populifolius), Sage-leaved Cistus (Cistus salvifolius), White Scrub-rockrose* (Halimium umbellatum), Alison Scrub-rockrose* (Halimium alyssoides), Strawberry Tree (Arbutus unedo), Bell Heather (Erica cinerea), Besom Heath (Erica scoparia), Tree Heath (Erica arborea), Heather (Calluna vulgaris), Wood Sage (Teucrium scorodonia), Cevennes Thyme* (Thymus nitens cebennis), Wild Madder (Rubia peregrine), Small-leaved Helleborine (Epipactis microphyla), French Lavender (Lavendula stoechas), Cornelian Cherry (Cornus mas), Laurustinus (Viburnum tinus)

Flora of limestone steppes (routes 1, 2, 4, 5, 14, 15, 18)

If one were forced to choose only one habitat to go plant hunting, we would advise to head to the dry limestone steppe-meadows of the Causses. They are home to the region's most distinctive flora and the profusion of wildflowers, particularly in spring, is amazing.

The two most emblematic species are the Cardabelle and the Feather Grass. The Cardabelle is a large, pale yellow, ground hugging and stemless thistle with a flower head of up to 15 cm across. Its "official" English name is Acanthus-leaved Carline Thistle, which is a literal translation of its scientific name, but not nearly as poetic as the French "Cardabelle". This plant is strongly interwoven in the local folklore and is commonly seen pinned on doors in the hamlets of the Causses (see text box on page 85).

The other emblematic plant, Common Feather Grass, is the typical grass of

The grasslands of the open Causse support a large range of wildflowers.

the Causses. Whilst grass is, to many, not much more than that green stuff that fills the space between flowers, Feather Grass is, by any standards, an attraction in itself. Feather Grass forms wavy seas of feathery, seedy fluff in June and July (the name is thus well chosen, but the French win again with their lyrical *Cheveaux d'Ange*, Angel's Hair).

The Feather Grass will make a visit to the Causse in early summer unforgettable, particularly in the mornings and evenings when the low sun sets the grass ablaze with colour.

A wide range of species accompanies the Feather Grass (and other grasses). In May and June, when the flora on the Causse is at its peak, there are dense mats of White and Hoary Rockroses, mixed with Wild Mignonette, Mountain Kidney-vetch, Montpellier Milk-vetch, White Flax (which is actually pink), Pink Bindweed, mixed with many other species. The limestone steppes are also rich in orchids (see page 95).

In the more fertile soils larger and deeper rooting plants like Lizard Orchid, Meadow Clary, False Sainfoin, Mediterranean Yellow-rattle, Round-headed Leek and masses of a subspecies of Crimson Clover (with pinkish-creamy flowers instead of its normal crimson colour) are found.

The local subspecies of the Alpine Aster flowers abundantly on the Causses in May.

At the other end of the spectrum, on very rocky soils, you will find rock plants, like those growing in the gorges (see page 76): Typical species to look for on these very thin soils are Alpine Mezereon and its pink relative, the Garland Flower together with Yellow Leek, Grass-leaved Buttercup, Mountain Kidney-vetch, Rock Soapwort and various pinks.

The altitude of the Causses, together with the porous bedrock, create continental conditions much like those in the steppes of Eastern Europe and, as a consequence, many eastern steppe flowers can be found on the Causse, adding to the attractiveness of this habitat. If you have visited the dry grasslands of eastern or south-eastern Europe, the Causses may remind you of these travels because of the drifts of Steppe Spurge*, the patches with Yellow Odontites and Goldilocks Aster, the sky blue flowers of Prostrate Speedwell and, occasionally, the impressive Ethiopian Sage.

There are also a handful of species, restricted to the Causse of the Massif Central, Cevennes Odontitis* *(Odontites cebennensis)*, Girard's Thrift* *(Armeria girardii)*, Grass-leaved Oxe-eye daisy* *(Leucanthemum graminifolium)* and two beautiful orchids: Aymonin's Orchid* *(Ophrys aymoninii)* and Aveyron Orchid* *(O. aveyronensis)*.

Typical plants of the Causses

Bastard-Toadflax (Thesium divaricatum), Buckler Mustard (Biscutella laevigata), Montpellier Pink (Dianthus monspessulanus), Wood Pink (Dianthus sylvestris), Rock Soapwort (Saponaria ocymoides), Grass-leaved Buttercup (Ranunculus gramineus), Yellow Pheasant's-eye (Adonis vernalis), Pheasant's-eye (Adonis annua), Large Pheasant's-eye (Adonis flammea), Pasqueflower (Pulsatilla vulgaris), Dark-red Pasqueflower* (Pulsatilla rubra), Hairy Sandwort* (Arenaria hispida), Flax-flowered Sandwort* (Minuartia capillacea), Girard´s thrift* (Armeria girardii), Cardabelle (Carlina acanthifolia), Alpine Aster (Aster alpinus cebennis), Snowy Mespilus (Amelanchier ovalis), Box (Buxus sempervirens), Sessile-leaved Broom* (Cytisophyllum sessilifolium), Woolly Cotoneaster (Cotoneaster nebrodensis), Horrid Hedgehog Broom* (Echinospartium horridum), Red Trefoil (Trifolium rubens), Bellflower Flax* (Linum campanulatum), Burnet Rose (Rosa pimpinelifolia), Steppe Spurge* (Euphorbia seguieriana), Kidney-vetch (Anthyllis vulneraria), Crimson Clover (Trifolium incarnatum), False Sainfoin (Vicia onobrychoides), Field Eryngo (Eryngium campestre), Buttercup Hare's ear* (Bupleurum ranunculoides), Cross-leaf Gentian (Gentiana cruciata), Trumpet Gentian (Gentiana clusii (rare)), Fringed Gentian (Gentianella ciliata), Swallow-wort (Vincetoxicum hirundinaria), Squinancywort (Asperula cynanchica), Blue Bugle (Ajuga genevensis), Wall Germander (Teucrium chamaedrys), Rouyan's Felty Germander* (Teucrium rouyanum), Golden Felty Germander (Teucrium aureum), Mountain Germander (Teucrium montanum), Hyssop-leaved Ironwort* (Sideritis hyssopifolia), Wild Marjoram (Origanum vulgare), Common Lavender (Lavandula angustifolia), Ethiopian Sage (Salvia aethiopis), Meadow Clary (Salvia pratensis), Prostrate Speedwell (Veronica prostrata), Crested Cow-wheat (Melampyrum cristatum), Crested Lousewort (Pedicularis comosa), Yellow Odontites (Odontites lutea), Common Globularia (Globuaria punctata), Mediterranean Yellow-rattle (Rhinanthus mediterraneus), Tuberous Valerian* (Valeriana tuberosa), Yellow-flowered Giant-scabious* (Cephalaria leucantha), Small Scabious (Scabiosa columbaria), Grass-leaved Oxe-eye Daisy* (Leucanthemum graminifolium), Causses Oxe-eye Daisy* (Leucanthemum subglaucum), Carline Thistle (Carlina vulgaris), Woolly Thistle (Cirsium eriophorum), Dwarf Thistle (Cirsium acaule), Spotted Knapweed (Centaurea maculosa), Carduncellus (Carduncellus mitissimus), Hairy Viper's-grass* (Scorzonera hirsuta), Purple Viper's-grass (Scorzonera purpurea), St. Bernhard's Lily (Anthericum liliago), Branched St. Bernhard's Lily (Anthericum ramosum), Blue Aphyllanthes (Aphyllanthus monspeliensis), Star-of-Bethlehem (Ornithogalum umbellatum), Autumn Squill (Scilla autumnalis), Round-headed Leek (Allium sphaerocephalon), Yellow Leek (Allium flavum), Crimean Iris (Iris lutescens (= I. chaemiris)), Hairy Melick (Melica ciliata), Feather Grass (Stipa pennata)

Thistle on the door

When travelling through the Grands Causses you cannot fail to see dried flower-heads pinned to the wooden doors. What is the flower and what is its significance?

The flower is the Acanthus-leaved Carline Thistle, *Carlina acanthifolia*, known locally to the French as the Cardabelle. It is a low-growing perennial which can reach a size of 50 cm across with a central flower-head of up to 15 cm. Like other thistles the leaves are very spiky and "white-felted" below. The plant thrives on limestone between 500m and 1800m and occurs regularly on the more open areas of the Causses. It flowers in July and August. (Even though the Cardabelle is pinned to doors, it is a rare and vulnerable species that should not be picked).

Historically the flower was collected by the peasants and cooked to make a much appreciated dish, quite similar to artichoke hearts (artichoke is a relative of the Cardabelle), and was also used to make jams as a treat for the children. The plant has also been used to treat wounds and ulcers, to aid quicker healing, and also as a cure for influenza, acne and eczema. There is even an old story telling about an angel that appeared before Charlemagne, king of the Franks, instructing him to use the Cardabelle to cure his soldiers of the plague.

But why are they attached to doors? There are several stories relating to this and each has its own merit. The first one is that it is for good luck and to ward off evil spirits. But this is perhaps too simple. Other accounts point towards the sophisticated hygrometric capabilities of the plant. The bracts close over the flower head when the air becomes damp or when it rains turning the Cardabelle into a natural barometer, enabling people to tell the weather just by looking at the plant on their door. Whether this is really the reason for this habit is doubtful, for, as a Frenchman cynically remarked to us, once you stick your head out of the door to see what the plant is doing, you will know from direct experience whether it is raining or not. The final possible reason is that the closing of the bracts over the flower head is a symbol of protection for the family within, during the harsh times on the plateaux.

The Acanthus-leaved Carline Thistle or Cardabelle is often attached to doors of houses and other buildings on the Causses.

Flora of the fields (routes 1, 2, 14)

Just a few decades ago, one could see cereal fields with more poppies than barley and fallow land with such a blur of blue Cornflowers and Larkspurs, red Poppies and white Chamomiles that they looked like scrambled French flags. These days are over, but on a more modest scale these 'weeds' of the fields – now endangered in many countries – are still fairly common in the Cévennes. The cereal fields in dolines and in fertile corners of the plateaux, particularly on the Causses, support a range of colourful annual and biannual wildflowers. The red of the common Poppy is dominant, but also present is the pink of the Corn-cockles and the blue of the Cornflowers, Venus's Looking-glass, Common, Eastern and Forking Larkspurs. If you look more carefully in the deeper undergrowth, you might find Greater Rock-Jasmine, Ground-pine and various Cornsalads.

Poppies and Corn Flowers colour the corn fields on the Causse. On the edges of such fields you'll find several rare species, like Greater Rock Jasmine and Venus' Looking-glass.

Typical plants of cereal fields

Ground-pine (Ajuga chamaepytis), Summer Adonis (Adonis aestivalis) Pheasant's-eye (A. annua), Orlaya (Orlaya grandiflora), Common Larkspur (Consolida ajacis), Forking Larkspur (C. regalis), Crowned* Cornsalad (Valerianella coronata), Broad-fruited Cornsalad (V. dentata), Narrow-fruited Cornsalad (V. rimosa), Field Gladiolus (Gladiolus italicus), Crested Cow-wheat (Melampyrum cristatum), Greater Rock-Jasmine (Androsace maxima)

Flora of the forests (routes 1, 2, 5, 7, 11, 13, 16, 17, 18)

Forests cover the largest part of the Cévennes, but since many of them are no more than a few decades old (see page 34) they are a bit dull from a botanical perspective. Nevertheless, there are some interesting species to be found in almost any woodland (it is the Cévennes after all!) and many in certain specific types of forest.

The flora of the pine plantations on the Causses has something of a split personality. Whilst unarguably dull in terms of plant diversity, the component parts of this flora could hardly be called uninteresting. Beneath the regimented trees a veritable feast of helleborines can be found – Red, Dark-red, Narrow-leaved, Broad-leaved and White – often garnished with equally interesting plants such as One-flowered Wintergreen and Yellow Bird's-nest.

When hunting wildflowers in woodlands, you'll notice again a big difference between those on the limestone and those on the acidic granite and schist soils.

Flora of woodlands on limestone
(routes 2, 11, 18 oak; routes 1, 2, 14, 16 pine)

Three types of woodlands are found on limestone soils: oak forests (mostly dominated by youngish Downy Oaks), Pine forests (mostly Scots and Austrian Pine) and Beech forest. The latter is rare on limestone and restricted mostly to the inaccessible, north-facing cliffs.

The young Downy Oak Woods in the limestone valleys have a fair number of interesting plants, but on the whole they support only a narrow selection of Causses plants, those thriving in semi-shaded conditions like Sessile-leaved Broom* *(Cytisophyllum sessilifolium)* and Bastard Balm. Although it is certainly not a punishment to walk these woods, older woodlands are

Liverleaf grows in abundance in the Oak woodlands on the Causse edges.

much more attractive. They are usually found in the gorges where forestry is difficult. Here, the list of plants grows to include Spanish Spiny Greenweed* *(Genista hispanica)*, Wood Calamint, Wild Basil, Large-flowered Self-heal and Yellow Foxglove, amongst others. Typical of these forests is Hepatica and an endemic species of the Cévennes, Cévennes Lungwort* *(Pulmonaria cebennis)*.

In contrast to the pine plantations on the Causses, the pine forests in the gorges of the Jonte and Tarn are spectacular places for botanists. Due to the open character of these woods, forest species mingle with species more often found on the Causse, in scrubland and on cliffs. Typical are those plants that thrive in the dense layer of moss and pine needles. In this layer, the limestone has leached to provide more acidic and moist conditions where one can find northern species like One-flowered and Green Wintergreens amidst mats of Bearberry and Wood Cow-wheat. The latter has yellow flowers under a striking hood of purple leaves. The little orchid Creeping Lady's Tresses can be very abundant here.

Blue Catananche is a conspicuous flower in July. Search for it on the Causse and in grassy patches in the gorges.

Typical plants of forests on limestone (either downy oak, pine or both)

Liverleaf (Hepatica nobilis), Yellow Anemone (Anemone ranunculoides), Common Columbine (Aquilegia vulgaris), Sessile-leaved Broom* (Cytisophyllum sessilifolium) Spanish Spiny Greenweed* (Genista hispanica), Spring Pea (Lathyrus vernus), Alpine Mezereon (Daphne alpina), Bearberry (Arctostaphylus uva-ursi), One-flowered Wintergreen (Moneses uniflora), Green Wintergreen (Pyrola chlorantha), Yellow Bird's-nest (Monotropa hypopytis), Yellow-wort (Blackstonia perfoliata), Cross-leaf Gentian (Gentiana cruciata) Wood Calamint (Calamintha sylvatica), Large-flowered Self-heal (Prunella grandiflora), Bastard Balm (Mellitis melissophyllum), Yellow Foxglove (Digitalis lutea), Wall Germander (Teucrium chamaedrys), Wood Cow-wheat (Melampyrum nemorosum), Cévennes Lungwort (Pulmonaria cebennis), Purple Gromwell (Lithospermum purpurocoerulea), Peach-leaved Bellflower (Campanula persicifolia), Nettle-leaved Bellflower (Campanula trachelium), Martagon Lily (Lilium martagon), Blue Aphyllanthes (Aphyllanthes monspeliensis), Wild Tulip (Tulipa australis), Angular Solomon's-seal (Polygonatum odoratum), Creeping Lady's-tresses (Goodyera repens), Dark-red Helleborine (Epipactis atrorubens), Red Helleborine (Cephalanthera rubra), White Helleborine (Cephalanthera damasonium), Narrow-leaved Helleborine (Cephalanthera longifolia)

Flora of woodlands on schist on granite (routes 7, 13, 17)

The moister schist soils and montane climates give the Oak and Beech woods of the eastern and central Cévennes a more temperate European character than the limestone forests in the west. Familiar forest plants such as Yellow Archangel, Foxglove, Early Dog-violet and May Lily can be found here.

The more interesting flowers you'll encounter in specific spots only, particularly there where the soil is slightly richer and moister and when the forest is older and more natural. Such sites can be found on the north and east slopes of the Lozère and locally on the south slope of the Aigoual. Where the canopy has not yet developed its shady parasol, the forest floor is carpeted with Yellow and Wood Anemones, Moschatel Flower, Solid and Bulbous Corydalis, Five-leaved and Seven-leaved Bitter-cresses, the delicate Herb-paris and several other species that are typical of temperate European forests. A little later, in May, large drifts of Wild Tulips are a magnificent sight in the woods (there is a good site for them on route 3).

The Martagon Lily is one of the most impressive wildflowers of the summer months, with stalks of flowers that grow up to a metre.

Later in the year, the streams and springs in these woods are lined (and not infrequently overgrown) with what the French call 'megaphorbes', large herbs that benefit from the nutrient-rich and permanently wet soils. The huge leaves of White Butterbur battle for light with the large yellow-flowered Austrian Leopardsbane, Marsh Hawk's-beard and the white umbels of Wild Angelica. Closer to the water, the cream-coloured hoods of Wolfs-

Two common forest flowers: Foxglove (left) and Yellow Bird's-nest (right). The first is found mostly on Schist soils, while Yellow Bird's-nest is found in any forest type, but particularly abundant in pine plantations on the Causses.

bane and the pink of Apple Mint are frequent. Locally, the pretty light-blue flowers of the endemic Cévennes Rock-cress grace the streamsides.

Higher upstream, the mountain version of streamside flora supports Alpine species like Napels' Monkshood* *(Aconitum napellus)*, Globeflower, Plumer's Sow-thistle* *(Cicerbita plumieri)*, Purple Lettuce, Adenostyles and Aconite-leaved Buttercup.

Typical plants of montane beech and oak forests over acidic soils

Purple Lettuce (Prenanthes purpurea), Alpine Leek (Allium victorialis), Yellow Archangel (Lamiastrum galeobdolon), Yellow Anemone (Anemone ranunculoides), Large-flowered Calamint (Calamintha grandiflora), Hollowroot (Corydalis cava), Bird-in-a-bush (Corydalis solida), Five-leaved Bitter-cress (Cardamine pentaphyllos), Seven-leaved Bitter-cress (Cardamine heptaphylla), Alpine Squill (Scilla bifolia), Wood Stitchwort (Stellaria nemorum), Ghost Orchid (Epipogium aphyllum), Common Twayblade (Listera ovata), Coralroot (Corallorhiza trifida), Martagon Lily (Lilium martagon), Dog's Mercury (Mercurialis perennis), Bird's-nest Orchid (Neottia nidus-avis), Herb-Paris (Paris quadrifolia), Early Dog Violet (Viola reichenbachiana), May Lily (Maianthemum bifolium), Hedge Woundwort (Stachys sylvatica), Foxglove (Digitalis purpurea), Yellow Foxglove (Digitalis lutea), Whorl-leaved Solomon's-seal (Polygonatum verticillatum)

Flora of the subalpine heathlands, meadows and bogs (routes 7, 8, 9)

The final habitats that you really ought to visit, if you are interested in wildflowers, are the heathlands, meadows and bogs that are dominant on the high slopes of the Cévennes. In this respect, the Mont Lozère is much richer than the Aigoual.

The scenery is wonderful, with fresh mountain meadows full of wildflowers and haphazardly strewn blocks of granite. The meadows are in full bloom from May to July-August. They start out with seas of Pheasant's-eye and Wild Daffodils, Marsh Marigold and Purple Crocus. On the highest slopes, the Spring Pasqueflower starts to flower just after snow-melt.

In May and June, when fields of Piorno Broom* *(Cytisus oromediterraneus)* are in full flower, Mont Lozère has even more to offer. The meadows are full with White and French Rampions, Bistort, Common Bugle and Spignel, Wild Pansy, Wood Storksbill and Globeflower. The drier areas are dominated by heathlands, with the bright-blue Perennial Sheep's-bit in abundance, locally accompanied by the darker blue flowers of Globe-headed Rampion and French Bellflower. Mountain Everlasting and Arnica are two species that have become rare in north-west Europe, but are, in suitable habitat, still abundant on the Lozère.

Subalpine meadows on the Mont Lozère.

92

Pheasant's-eye Daffodil (left) colours the meadows on the Lozère white. The Grass-leaved Plantain (*Plantago holosteum*; right) grows in dry heathlands and between schist rocks.

The highest, coldest and most windswept slopes are relatively poor in species, but Alpine Clover and Alpine Lady's Mantle give them a distinct Alpine touch. A rarity is the superb St. Bruno's (or Paradise) Lily.

The springs and bogs and little mountain streams at this altitude support yet another botanical attraction. From the edges of these wetlands (they shouldn't be entered because they are very fragile) it is often possible to spot Round-leaved Sundew, Marsh Cinquefoil, Marsh Violet, Heath Spotted Orchid, Starry Saxifrage and Marsh Valerian. A rarity here is the Bog Asphodel, a beautiful yellow species of lily that occurs in many Atlantic bogs in north-west Europe. Here on the Mont Lozère, with its feet almost in the Mediterranean, it is at the absolute south-east extremity of its range.

Typical plants of the high slopes of the Mont Lozère

Spring Pasqueflower (Pulsatilla vernalis), Pheasant's-eye Daffodil (Narcissus poeticus), Wild Daffodil (Narcissus pseudonarcissus), Marsh Marigold (Caltha palustris), Purple Crocus (Crocus vernus), Piorno Broom* (Cytisus oromediter-raneus), White Rampion (Phyteuma spicatum), French Rampion (Phyteuma gal-licum), Bistort (Persicaria bistorta), Common Bugle (Ajuga reptans), Spignel (Meum athamanthicum), Wild Pansy (Viola tricolor), Wood Storkbill (Gera-nium sylvaticum), Globeflower (Trollius europeus), Marsh Violet (Viola palus-tris), Water Forget-me-not (Myosotis scorpioides), White False-helleborine (Veratrum album), Yellow Gentian (Gentiana lutea), Marsh Gentian (Gentiana pneumonanthe), Heath Spotted Orchid (Dactylorhiza maculata), Small-white Orchid (Pseudorchis albida), Wild Tulip (Tulipa australis), Maiden Pink (Dian-thus deltoides), Marsh Lousewort (Pedicularis palustris)

High slopes

Alpine Clover (Trifolium alpinum), Globe-headed Rampion (Phyteuma hemi-sphaericum), Fleabane-leaved Hawk's-beard* (Crepis conyzifolia), Perrenial Sheep's-bit (Jasione laevis), Mountain Everlasting (Antennaria dioica), Golden cinquefoil (Potentilla aurea), Seguier's Pink (Dianthus seguieri), Arnica (Arnica montana), St. Bruno's Lily (Paradisea liliastrum), Hairy greenweed (Genista pilosa), Spiny Greenweed (Genista anglica), Winged Greenweed (Chamaespar-tium saggitale), Heath Dog Violet (Viola canina), Lesser Yellow-rattle (Rhinan-thus minor), French Bellflower (Campanula recta), Alpine Lady's-mantle (Al-chemilla alpina)

Bogs

Whorled Caraway (Carum verticillatum), Bog Asphodel (Narthecium ossifra-gum), Grass-of-Parnassus (Parnassia palustris), Marsh Violet (Viola palustris), Marsh Valerian (Valeriana dioica), Marsh Cinquefoil (Potentilla palustris), Py-renean Angelica (Selinum pyrenaeum), Round-leaved Sundew (Drosera rotundi-folia), Small Cranberry (Vaccinium microcarpum), Heath Milkwort (Polygala serpyllifolia)

Stream borders

Water Avens (Geum rivale), Large Bitter-cress (Cardamine amara), Opposite-leaved Golden-saxifrage (Chrysosplenium oppositifolium), Alpine Willowherb (Epilobium anagallidifolium), Starry Saxifrage (Saxifraga stellaris)

Orchids

Routes 2, 4, 14, 16 and 18 are particularly good for hunting orchids. On page 220 a full list of orchids is given, including their habitat and flowering time.

The Cévennes and Grands Causses rank amongst the best regions for finding wild orchids in the whole of Europe. Due to many recent (and ongoing) changes in taxonomy, it is not possible to give an objective number for the orchid species present in the region. New analytical techniques suggest that some former "species" are in reality an aggregate of different, yet highly similar species, and hence their exact distribution is not known. This makes enumerating, let alone identifying, the orchids of the Cévennes a bit tricky, but according to our count, there are no less than 61 species present (see table on page 220). With this number, the Cévennes and Grands Causses region can compete with such classic orchid haunts as Cyprus, Crete, South Italy and South-west Turkey. In the Cévennes it is not just the large amount of species that attracts, but also the fact that many of them occur in such profusion that you could almost consider them weeds. It seems, therefore, strange that the Cévennes, nearby and accessible for many enthusiasts, are not more widely known amongst orchid aficionados.

This being said, orchids are far from evenly distributed over the region. In some areas orchids abound, whilst in others they are almost absent. Most species are confined to the limestone areas so, consequently, the Causses and the peripheral hills in the south and east (see map on page 22) should be your first targets if orchids are your interest. The schist and the granite landscapes are a lot less well endowed with orchids, and in the areas of broom scrub and heathlands, any search for them will be in vain. Only in very specific habitats will you find orchids on granite and schist, but most of these species are quite special.

Orchids on the Causses

The orchid flora of the Causses is a mixture of species typical of Europe's temperate regions and those with a Mediterranean range with a few representatives of the colder regions. The first group is particularly well represented on the Causses. Smaller and larger drifts of orchids occur all over the Causses and typically include Military, Lady, Monkey, Burnt, Man, Green-winged, Early-purple, Elder-flowered, Fragrant, Pyramidal and Lizard Orchids. Also fairly common are Greater and Lesser Butterfly Orchids, Greater Twayblade and Frog Orchid. It would be exceptional to find all these species together, but if you explore the Causses for a day or two in the right season (particularly on some of the described routes) you should be able to find most of the abovementioned. When visiting the open pine woods on the Causses, you should be able to add Red, White, Narrow-leaved, Narrow-lipped, Müller's and Dark-red Helleborines, Bird's-nest and Violet Bird's-nest Orchids and Creeping Lady's-tresses to the list (in appropriate season; see again table on page 220). The superb Lady's-Slipper Orchid grows on a few very-hard-to-reach localities on north-facing slopes of the Gorge du Tarn.

Pyramidal Orchid (opposite page) is a common species on the Causses. Red Helleborine grows in open pinewoods on limestone. It is less common than White and Narrow-leaved Helleborines.

96

Monkey Orchid is a frequent Causse orchid.

Several species of the Bee Orchid genus *(Ophrys)* can be found on the open limestone grasslands. Among them are the Bee Orchid and the Fly Orchid, the latter of which is an easily overlooked beauty. Once you have developed an eye for it, you will find the Fly Orchid growing quite often among the more flashy species like Military and Monkey Orchids.

For the true orchid enthusiast, three species (all of the Bee Orchid genus) make the trip to the Cévennes particularly worthwhile, because they are endemic (restricted) to very small ranges centred on the Cévennes. These are the Small Spider Orchid* *(O. araneola)*, Aymonin's Orchid* *(O. aymoninii)* and the Aveyron Orchid* *(O, aveyronensis)*. The Small Spider Orchid* is unique to Central France and the border areas with Spain and Italy. It is recognisable by its long, many-flowered stalks of small, pale flowers. The Aymonin's Orchid* with its exquisite flowers has an even smaller range which doesn't extend beyond the borders of the Grands Causses region. Its range is centred around the Causse Méjean and Causse Noir, where it grows on the thin soil of the open limestone grasslands, often accompanied by other, more widespread orchids. Aymonin's Orchid* (photo on back cover) is readily recognised as a type of Fly Orchid with a broader, yellow-fringed lip and green petals. Just as unique is the Aveyron Orchid* (p. 206) which is, as the name implies, restricted to the district of Aveyron in the southern Causses. In fact, this pink-flowered species of the Spider Orchid group is known only from a limited number of warm localities on the Causse du Larzac. These three species can be found – in season – along the routes set out in this book on the Causse du Larzac and Causse Méjean.

The further south you go on the Causses, the more Mediterranean species can be found. The Causse Larzac, in particular, offers many sites with southern species like Woodcock, Yellow Bee and Early Spider Orchids and, more rarely, Furrowed Orchid* *(O. sulcata*, a local species of the Dull Bee Orchid). Of these Mediterranean species, the group of Spider Orchids is best represented. Apart from the afore-mentioned, you can find – in season – Black Spider Orchid and Passion-tide Orchid* *(O. passionis)*. The latter, formerly not considered distinct from Early Spider Orchid, is the most abundant. The Black Spider Orchid* *(O. atrata)* is probably confined to the foothills of the Cévennes with their Mediterranean climate.

The southern and eastern slopes

The hot, limestone hills on the edges of the Cévennes form another exciting area to search for orchids. Their position, well within the influence of the Mediterranean climate, make it possible to find species that are absent in the Cévennes proper. However, the peripheral hills are not well researched, so what is actually present, remains something of a guess. Bee, Woodcock, Yellow Bee, Early Spider, Small Spider and Lizard Orchids are certainly present, but Giant Orchid is likely to be found here too. The rare Large-flowered Orchid* (*O. magniflora*; a relative of Berteloni's Orchid) is reputed to grow in the limestone hills on the southern and western rim of the Cévennes. Our searches for it have been in vain, but you might be more lucky. The rare and beautiful Pink Butterfly orchid occurs only near Saint Jean de Luzencon.

Orchids on schist and granite

Most of the above-mentioned limestone species flower in April and May, with Lizard and Pyramidal Orchids, Red, Dark-red and Narrow-lipped Hel-

leborines and Creeping Lady's-tresses flowering in June and into July. However, there are a number of species that flower in midsummer. The majority of these are not found on limestone, but on schists and granite soils. In the cool mountain beechwoods, on acidic soils, you can search for the diminutive Coral-root and Ghost Orchid, two species that grow in wet, rotting wood and leaf litter.

The river valleys in the Schist Cévennes form another haunt with rare orchids. In moist rock cracks along these rivers in July you might find the tiny Summer Lady's-tresses, a rapidly declining species of the Mediterranean and Atlantic regions. Earlier in the year these same areas are the place to find Tongue Orchid, another truly Mediterranean species that reaches its northern limits in the Cévennes.

Orchid enthusiasts with some more time to spend could consider some visits to other areas nearby, such as the Vercors for Alpine species or the Provence for some rare Mediterranean orchids.

The Fly Orchid is one of the prettiest orchids. It is a very small flower and is frequent in low herby grasslands.

98

Mammals

There are no particular routes suitable for watching mammals. Many of the large species are very shy due to hunting and it takes a considerable amount of luck to find them. Driving the minor roads at dusk increases your chances somewhat. In contrast, Beaver watching is relatively easy. The area around Florac is very good, but all river gorges have their good spots. Ask at riverside campsites. The Jonte and Tarn gorges are superb for bats.

One would expect that the Cévennes and Grands Causses, being a wild, remote and forested region, support a rich mammal fauna including all the large herbivores and carnivores. Unfortunately, this is not the case. The centuries-long hunting pressure and, in particular, the disappearance of most of the woodlands (see history section on page 56) has banished the large carnivores like Brown Bear, Lynx and Wolf from the Cévennes and Causses.

The bigger game animals, like Red Deer, Fallow Deer, Roe Deer, Wild Boar and Mouflon, all occur in the Cévennes, but – except for Roe Deer and Wild Boar – have been (re)introduced and managed for hunting. The Cévennes National Park is the only French National Park where people live and where human culture and customs are an integrated part of the

Beavers in the Tarn. Beavers strip the bark of willow twigs at high speed. These white, stripped branches are a clear sign of the presence of Beavers in a river.

conservation goals and strategies. This also means that hunting is allowed in many places, which influences the large mammal fauna.

Except for the native Wild Boar and Roe Deer, the large mammals occur quite locally in the areas where they were originally introduced. For example, several large herds of Mouflons – the wild mountain sheep of Corsica – occur on the south slopes of the Mont Aigoual and only a few other places. After its disappearance from the Cévennes in the 18th century, Red Deer was reintroduced in 1961 in the Massif Aigoual, near Meyrueis, and in the Massif de Goulet where they are now quite common.

On a 3000 ha terrain of particularly open and steppe-like grassland on the Causse Méjean (route 2), a herd of Przewalski Horses is kept in semi-natural conditions. The horses are being bred for a reintroduction project in Mongolia, their native country (see text box on page 69).

In contrast to the ungulates, the many species of smaller mammals are not managed. There are Otters in the larger rivers, Foxes, Badgers, Pine and Stone Marten in the woodlands and the beautiful, but shy, Genet in the scrub and woodlands in the warm, lower regions. In other words, most of the smaller carnivores that were originally present in the Cévennes, are still flourishing. Data is hard to come by for these secretive species, but it can only be that their populations have grown with the increase of woodland during the 20th century.

The best chance you have to catch a glimpse of these secretive species is by slowly biking or driving the minor roads – through the appropriate habitat of course – at dusk or dawn and see what appears in your headlights.

Beavers

Much easier to spot is the Beaver. This large, river-dwelling rodent was nearly eradicated for its fur and for its tendency to meddle with the river courses. In France, the Beaver was nearly extinct by the end of the 19th century, with remaining populations only in the lower reaches of the Rhone. After its protection in 1909, the Beaver recolonised the eastern section of the Cévennes. The National Park board speeded up the process by reintroducing the Beaver to the Atlantic watershed (the rivers flowing westwards) in 1977. They were first introduced into the Tarn and are now abundant in the rivers Tarn, Jonte and Dourbie.

In contrast to those found in the lowland regions of Europe, Beavers are not very shy in the Cévennes. They are, in fact, quite easy to spot. Many campsites at the banks of the west-flowing rivers have a Beaver colony nearby (e.g. those on the Tarn at Florac) where beaver-watching has become a popular evening outing. You can find Beavers yourself by looking for debarked branches in the water (see photo on page 98), but the easiest way to catch a glimpse of a beaver is by asking your neighbours at one of the campsites along the Tarn, Jonte or Dourbie 'pour observer le Castor'. The best beaver-watching sites are often very well known locally.

Bats

With all the caves, cracks, cliffs and abandoned houses, there is an abundance of bats. Lesser and Greater Horseshoe Bats are common in the gorges, often forming sizable colonies. Also, Daubenton's Bat is very common. It is a typical 'water bat', always flying low over rivers to catch insects in mid-air or from the water surface. A large number of other bats, including Natterer's, Leisler's, Whiskered, Notch-eared, Long-eared and Grey Long-eared Bats and Kuhl's, Savi's and Common Pipistrelles, occur widely in the Cévennes. They hunt in a variety of habitats like open woodland, forest edges and rural areas. Lesser and Greater Mouse-eared Bats occur primarily on the Causses. There are several other species of bat also present in the region, but they are rare.

Furthermore, one can find Hares, Rabbits (although quite rare), Hedgehogs, Red Squirrels, Ermines, Weasels, Polecats, Edible Dormice and a variety of mice and shrews in the Cévennes and Grands Causses.

The Roe Deer is a widespread but shy animal in the Cévennes.

Birds

The best birdwatching is on Causse Méjean (routes 2, 14, 15 and 16) and Causses Blandas (route 6). Birds which prefer higher altitudes are best on the Mont Lozère (particularly route 9). To locate birds of rocky gorges and Mediterranean scrubland, try route 2, 11 and 19. The best route to watch vultures from up close is route 2. A species by species birdwatching guide is given on page 224.

Nightjars breed in the open forest and scrub on the Causse. Their typical churring can be heard on warm spring evenings in May and June. They sometimes come down in the valleys to hawk over the rivers.

The Cévennes and the Grands Causses are, perhaps, the most underwatched natural areas in France for birds. Little has been written about them, even though they have much to offer the birdwatcher; in quantity they may not be top notch, but in quality and diversity they are quite impressive. Due to the altitude and steep terrain, the region attracts several mountain species whilst its proximity to the Mediterranean Sea (only 70 kms as the vulture flies) means that there are also some Mediterranean species that reach their northern limits in the Cévennes. Mostly though, you will encounter the typical woodland and scrubland birds of the French countryside.

Most species are familiar to visitors from north-west Europe, but there are also some southern species, like Crag Martin, Bonelli's and Melodious Warblers and Short-toed Eagle, all of which occur in good numbers. Honey Buzzard, Cirl Bunting and Red-backed Shrike are abundant in, for them, suitable habitat.

The really good birdwatching is limited to a few particular sites in the area. These are the Causses (particularly Méjean, Larzac and Blandas) for steppe birds, the gorges (particularly Tarn, Jonte and Vis) for Mediterranean birds and the Mont Lozère for birds of the high mountains. Below, we describe these sites in more detail.

Birds of the Causse Méjean

The Méjean is arguably the most interesting of the Causses for birdwatching. It sees the highest density of raptors in the region. The impressive vultures with their huge wingspans are the most visible. They breed in the Gorge de la Jonte (see text box on page 104) and since the Méjean is closest to the colony, it is almost impossible to scan the skies here without detecting a vulture of some sort. Griffon Vultures are the default 'species', while Black Vultures tend to be seen in ones and twos among large foraging parties of Griffon Vultures. The Black Vulture is the largest bird of prey here and in fact, after the Condors, the largest bird of prey in the world. It is the only European vulture to nest in trees, though it is not the best of nest builders. Some of the nests lean at alarming angles and it is surprising that they don't fall out of the trees.

The third Vulture of the Cévennes, the Egyptian Vulture, is much rarer than the other two, with only two regular breeding pairs in the area. The Cévennes is right at the edge of its breeding range and they probably never were abundant in the region.

The Black Vulture was introduced to the Grands Causses in 1992.

The Hen Harrier breeds in low numbers in all open country of the region, from limestone grasslands to heathlands.

After the vultures, the most typical bird of prey in the Causses is the Short-toed Eagle. There are few places in Europe where this species is as common as it is here. Short-toed Eagles spend much of their time hovering or soaring across these open plains looking for their favourite food – snakes. Golden Eagles were once common as well, but have become very scarce since the introduction of myxomatosis disease decimated the population of their favourite food, the rabbit. In 1945 there were over 50 pairs breeding in the Lozère region whereas today there are only five, making it a hard-to-find rarity (in contrast, populations are stable in the Pyrenees and Alps because there the abundant Marmots are the main prey). Both Hen and Montagu's Harriers are frequently seen quartering the grasslands across the Méjean, looking for small rodents and young birds. The Hen Harrier is the scarcer of the two, being outnumbered by the Montagu's by about 4 to 1.

The open character of the Causse Méjean makes it attractive for some birds of Mediterranean and steppe-like terrain that are not found anywhere else in the Cévennes, except for the Blandas and Larzarc Causses. Hoopoes are more numerous here than elsewhere in the Cévennes, and Tawny Pipits are frequent in the thin grass swards. The Black-eared Wheatear, a Mediterranean species, can be found on rocky slopes around the small village of Nivoliers and is rarely found elsewhere. Only about a dozen pairs are recorded annually. The Northern Wheatear is much more numerous and in the strong light this species sometimes seems so pale that you have to be careful not to confuse it with Black-eared Wheatear. Short-toed Lark – although very rare – is another specialty that can be found around the villages of Hures, Drigas and Villaret. Occasionally, it turns up at lavognes to drink.

The decline and successful recovery of the Griffon Vulture

The vultures are, once again, the most spectacular of the Cévennes birds. "Once again" because, although the Griffon has been present in the region as far back as records go, they became extinct in 1941. They were hunted in the name of sport and it was not unusual to see hunters carrying several birds on a pole, suspended unceremoniously by their feet. Shepherds too would shoot vultures, thinking that they might kill their lambs. This is not the case, vultures only eat carrion.

However, it was not solely hunting that drove the vultures into extinction. Quite unexpectedly, it was the unintended result of developments that followed the discoveries of the famous speliologists (cavers) Louis Armand and Eduard Martell that did the vultures in. They discovered the now famous Aven Armand, an enormous and spectacular swallow-hole, large enough to hold Notre Dame Cathedral. Martell and Armand started to experiment with coloured dyes to see what happened with the rainwater that entered the cave. They discovered that the water didn't just disappear but emerged from the cliffs further down the valley to continue its way into the Jonte. However, local farmers had long tipped the bodies of dead animals down the opening of the Aven. It was feared that the rainwater that flowed past the cadavers contaminated the river water on which towns and villages downstream relied!

It did not take long for the government to take action on this unhygienic situation. In the early 1900s they formulated a decree that all dead animals had to be buried to prevent any chance of disease. This effectively removed most of the natural food for the vultures, which created a famine and this caused the demise of the vulture population.

A French environmentalist, Michel Terrasse, together with his brother and two friends, set up the 'Fund for the Protection of Birds of Prey', heralding the reintroduction of the vultures. In 1970, four young sick or injured birds were brought in from Spain and, once recovered, released into the wild. The results were disappointing to say the least. The first was shot by a hunter, the second was electrocuted on an electricity pylon and the other two, after roaming around for a while, were never seen again. Only eight months after the initial release, the Jonte cliffs were as deserted as before.

A new approach was needed and it was decided to release only adult birds, as they were less likely to roam from the release site once they had paired (Griffon Vultures pair for life). Larger aviaries were built and more birds brought in from Spain, but this time they were allowed to reach maturity and get used to their surroundings before release. Once the first birds were released, in December 1981, food in the form of sheep carcases were left out close to the aviaries in order to keep the birds fed. This worked perfectly and the first chick was born in 1982. Before the second release, the team spent much time discussing the reintroduction project with the local shepherds and hunters. This resulted in not a single bird being shot since the start of the scheme. The second hurdle was to gain an exemption from the devastating law regarding the burying of carcases on the Grands Causses and this was also achieved prior to the release. Between 1981 and 1986 a total of 61 vultures were released.

Griffon Vulture.

Following the success of the Griffon Vulture reintroduction, Michel decided to adopt a similar scheme to bring back Black Vultures to the Grands Causses after an absence of about 120 years. Unlike the colonial rock-dwelling Griffon Vulture, the Black Vulture breeds in single pairs in the crowns of large trees. Towards the end of the 19th century most of the woodlands across the Causses and Cévennes were felled, mainly to fuel the glass factories in Millau, resulting in the demise of the Black Vulture due to the lack of suitable breeding sites. Using only young Black Vultures, which had been bred in captivity, Michel Terasse released the first birds in 1992 after a four-year period of acclimatisation in aviaries. The project has indeed enjoyed great success leading to 15 breeding pairs by 2007. This includes one pair, which nested in the Jonte Gorge immediately opposite the vulture exhibition centre, affording fantastic views for the visitors. The population has grown to become one of the largest in Europe, with over 60 individuals present in 2007. The third Vulture, the Egyptian vulture, had also disappeared from the region in the mid 1950s, but it returned naturally in 1986. However, with the species declining sharply in its main European stronghold, Spain, the small Cévennes population is becoming more important and reason for firm conservation measures.

What once was the crown on Méjean's birdlife, the Little Bustard, is sadly no longer present. It went with a sense of drama – the last known calling male was killed in 1996 by a Peregrine Falcon just as a camera crew was preparing to film it for a promotional (!) video.

Still fairly common on Causse Méjean is the Stone Curlew. This bird of the steppe and other dry open vegetations is frustrat-

Stone Curlew (top) and Tawny Pipit (bottom), two species of steppes and other dry, open country that thrive on the Causses.

ingly difficult to see due to its cryptic plumage. Your best chance to locate it is to listen for its eerie curlew-like calls, particularly just before dusk.

The Méjean is a good place for shrikes too. The Causses are exactly at the point where the ranges of Great Grey Shrike and Southern Grey Shrike overlap so it is possible to see both in the same vicinity. Southern Grey Shrike is the rarer of the two, although Great Grey Shrike is not a common species either. Much more numerous is the Red-backed Shrike which spends much of its time perched on top of Box or Juniper bushes, making it one of the easier birds to observe. With a little luck you'll find a shrike's 'larder', where the bird has impaled a luckless beetle or other insect. This habit not only creates a larder for times when food is less easily caught (e.g. during poor weather), but also helps the weak-footed shrikes dismember their prey and may even attract mates – males with bigger larders are more successful in attracting females.

Around the hamlets of Hures, Drigas and Le Buffre it is possible to find small colonies of Rock Sparrows. This Mediterranean bird is generally rare, and the Cévennes is one of the better places to find it. Buntings abound as well. Yellowhammer, Cirl and Corn Buntings can be found almost anywhere throughout the Causses, but Ortolan and Rock Buntings are localised and quite rare.

Large numbers of rock piles and karst rocks across this terrain provide perfect lookouts for Little Owl and the beautiful Rock Thrush. At such sites you will have little difficulty seeing the playful groups of Red-billed Choughs. They breed in the ruins and roam the Causses for large insects and lizards. This is also the place to look for Rock Bunting.

The Black-eared Wheatear is a Mediterranean bird species that reaches its northern limits on the Causse. It is a rarity here, most likely to be found on Causse Blandas (route 6) and in the southeastern part of Méjean (route 2).

Birds of the Causse Larzac and Blandas

Causse Larzac, with Blandas as a western 'sub'causse, is the largest of all the Causses.

The large military camp on Causse Larzac is reputed to house the last remaining population of Little Bustard. However, no one has heard or seen them for years and it is highly unlikely that they are still here.

The dry steppe landscape of the Causse Méjean also characterises Larzac and Blandas, but the latter also has large areas of scrub and woodlands. Larzac seems to lack the most typical Méjean species like Short-toed Lark and Black-eared Wheatear, but houses good populations of Short-toed Eagles, Common Buzzard, Honey Buzzard, Montagu's Harrier and Hen

Harrier, as well as Stone Curlew and Tawny Pipit. In addition, Larzac – more particularly the area to the south of la Couvertoirade (route 4) – is home to one of the few populations of Dartford Warblers in the region. In both 2002 and 2003, Bee-eaters bred nearby and could often be seen taking insects above the car park of La Couvertoirade.

Nightjars occur on all the Causses and in the heathlands of the Cévennes, but the Larzac seems to have the highest numbers. They regularly come down to drink at the lavognes just before dusk, creating ideal conditions for observing this bird (see route 4).

Stonechat

Causse Blandas, one of the region's best birdwatching sites, is the Causse with the most outspoken Mediterranean character. There are plenty of Rock Thrush, Stone Curlew, Tawny Pipit, Ortolan Bunting, Quail, Hoopoe and Subalpine and Orphean Warblers. The Orphean Warbler's population has increased over the last decade and its song, a series of short sweet phrases which can be likened to that of Ring Ouzel, is no longer an unusual sound on the Causse. When not in song, they are very skulking birds and you will need some patience to get good views. On the edges of the Causse, you can see Golden Eagle and Alpine Swift, which breed in the gorge.

Blandas is best known for its shrikes, however. Apart from Red-backed Shrike (common) and Southern Grey Shrike (scarce) it has a remarkable density of Woodchat Shrikes. This beautiful Mediterranean species is difficult to locate further north, but on Blandas, a careful observer doing route 6 of this book can easily find a dozen or more birds perched on the shrubs.

Birds of gorges and river valleys

The Gorges and river valleys that separate the Causses support a number of cliff and scrubland birds that makes them an attractive terrain for bird-watchers. Cliffs are not an easy terrain to cover. The tops of the cliffs are often better places to watch birds than the bottom. Crag Martins and Alpine Swifts fly around these cliffs constantly and it is the best place to wait and get good views. Likewise, vultures may pass by at the proverbial arm's length and Golden Eagles too soar by using the warm air that rises up at the cliff's edges. With some luck, you might also spot a Peregrine Falcon.

The Blue Rock Thrush is largely confined to these high cliffs and it is worth looking closely at any small bird on the top of a rocky outcrop. Being at the edge of its Mediterranean range, the Blue Rock Thrush is a rare bird in the region. Along routes 2, 6 and 19 you pass some reputed sites for this species, but also at the "Point Sublime", overlooking the Tarn Gorge north of Les Vignes, you have a good chance (and should you fail, the Alpine Swifts that pass a few feet overhead make the journey up worthwhile).

The rivers at the bottom of the gorges are home to several other species, such as Grey Wagtails and Dippers. Kingfishers dash along the larger rivers such as the Tarn which has suitable banks in which to nest. There are not many marsh birds around, the only duck species being the Mallard and the only heron being the Grey. Common Sandpiper is a scarce breeder, only found in the Tarn Gorge and on a few occasions the Little Ringed Plover have tried to nest on gravel banks in the Tarn but, the stream of passing canoes on the Tarn in summer disturbed them.

Blue Rock Thrush is another Mediterranean species rare in this region. You may find it on rocky terrain in the gorges, particularly in the higher regions of the southern Causses.

The remote wooded gorges are excellent nesting places for Buzzards and Short-toed Eagles. This is also the favourite haunt of Black Woodpeckers although they are more often heard than seen. There are many Tawny Owls to be heard at night and a few Long-eared Owls in more remote locations. The bird everyone hopes to see is, of course, the Eagle Owl which inhabits inaccessible caves and ledges high up in the Tarn and Jonte Gorges. The woodlands themselves hold many smaller species including Green and Great-spotted Woodpeckers, Nuthatch, Short-toed Treecreeper, Bonelli's Warbler, Firecrest and Crested Tit along with many more common species.

Birds of the Mont Aigoual

The densely wooded Mont Aigoual hosts the typical birds of coniferous forests and beech woods. You can encounter Coal Tit, Chaffinch, Bonelli's Warbler, Crested Tit, Great Spotted Woodpecker, Nuthatch, Eurasian and Short-toed Treecreepers, Firecrest and Goldcrest, but not a whole lot more.

The few areas with mature(ish) trees are suitable for Black Woodpeckers. In the older conifers you may come across the Common Crossbill, with its beak specially adapted for prising seeds from the pine cones. It nests as early as March and can often be seen in early May flying in family groups. The largee trees are also perfect nesting places for Goshawk. They prefer to do their hunting along forest rides and firebreaks. Honey Buzzards are also quite abundant here (as elsewhere in the Cévennes).

The avian specialty of the Aigoual is the Tengmalm's Owl. This typical

Crested Tit is found in the pinewoods, particularly in the gorges.

northern species (hence its other name, Boreal Owl) has one of its few French strongholds on the Aigoual – other sites are the Pyrenees and the Alps. No more than 20 pairs are believed to nest on the Aigoual where they use deserted Black Woodpecker holes to build their nests. Their courtship usually ends in April making it difficult to locate them after this time as they become almost completely silent in their habits.

Coal Tit is a bird of the Mont Lozère and Mont Aigoual.

Birds of the Mont Lozère

In contrast to the Aigoual, the Lozère has a large area of open, subalpine meadows and heathlands and therefore supports several birds you won't find on the Aigoual. The relatively level – as far as mountains go – highlands are ideal for such species as Skylark and Northern Wheatear, the latter conspicuously showing off its white rump as it flies away.

The Lozère's attraction lies in the occurrence of several Alpine birds, which occur together with the more widespread species of heathlands and forests. The two types of pipit that inhabit these slopes form a good example. The abundant Meadow Pipit is a widespread species in Europe, whereas the rarer Water Pipit typically breeds in pastures above the tree limit. Another Alpine species only found on the Lozère is the Ring Ouzel which resembles the more familiar Blackbird with a white crescent splashed on its breast. It nests in the wooded parts of the slopes but can

often be seen feeding in the open areas.

The highlight of the Lozère is the Citril Finch because it is confined to a very small range that extends from the Pyrenees to the Alps and Vosges and occurs nowhere else in the world (the Sardinian and Corsican populations are now believed to be a different species). This small bright-green bird is not hard to track down. It is usually seen in pairs or small groups in the meadows and around the hamlets of the Lozère. Other species that can be encountered on the Lozère are Common Crossbill, Linnet, Stonechat, Hen and Montagu's Harrier.

The Montagne des Bougès, just to the south of the Mont Lozére was the location of a reintroduction scheme involving the Capercaillie, Europes's largest member of the grouse family. Over 600 Capercaillies were released between 1978 and 1993 in the Scots Pine plantations. Unfortunately, the new population does not fare well. Several consecutive cold and wet springs resulted in low breeding success. But the main cause for the difficulties in the re-establishment of the Capercaillie is the explosion of the wild boar population since the 1990s, as these animals eat the Capercaillie eggs. So, in spite of the numbers released, it is thought that there are probably fewer than 80 birds left.

Birds in winter

With temperatures reaching -12°C, winter can be quite severe in the Cévennes. Only the hardiest birds stay in the area. They are joined by species from the north (including Merlin, Redwing and Fieldfare) and from the high slopes of the Alps (such as Alpine Accentor and Wallcreeper).

Merlin has been recorded on a regular basis as a scarce winter visitor, and has been mostly seen on Causse Méjean. Redwing and Fieldfare move into the south of Europe during winter in search of food and can be found almost anywhere. Alpine Accentors have become much more common in recent winters. Perhaps the most sought-after bird in Europe, by most British birdwatchers at least, is the Wallcreeper. This bird breeds in near-inaccessible places on sheer rock cliffs high above the treeline in the Alps and Pyrenees. The best time for seeing Wallcreepers is in winter when they move down to lower altitudes. This is when they can be found in the Tarn, Jonte and Dourbie Gorges (see page 228 for the best location).

The Citril Finch (opposite page) is found only in a few mountainous areas in western Europe but nowhere else in the world. It is fairly common on the Mont Lozère.

Bird list

Under 'specialities' the birds are listed that are rare or localised in the rest of Europe and the birds that are very abundant and easy to find in the Cévennes and thus typical for the region.

a = abundant, l = locally present, r = rare, vr = very rare or difficult to spot, i = introduced, w = winter only

Specialities Griffon Vulture (i), Black Vulture (i), Short-toed Eagle (a), Honey Buzzard (a), Stone Curlew (l), Tengmalm's Owl (l), Crag Martin (a), Alpine Swift (l), Tawny Pipit (a), Black-eared Wheatear (vr), Rock Thrush (l), Ring Ouzel (r), Melodious Warbler (a), Subalpine Warbler (l), Bonelli's Warbler (a), Red-billed Chough (l), Rock Sparrow (l), Citril Finch (l), Ortolan Bunting (l), Cirl Bunting (a), Alpine Accentor (w), Wallcreeper (w)

Others Egyptian Vulture (r), Golden Eagle (r), Red Kite (r), Black Kite (l), Hen Harrier (r), Montagu's Harrier (r), Peregrine (r), Quail (l), Red-legged Partridge (a), Turtle Dove (l), Kingfisher (l), Hoopoe (l), Bee-eater (vr), Black Woodpecker (l), Lesser Spotted Woodpecker (r), Wryneck (l), Crested Lark (r), Skylark (a), Woodlark (a), Eagle Owl (r), Scops Owl (r), Nightjar (l), Red-backed Shrike (a), Woodchat Shrike (l), Great Grey Shrike (r), Southern Grey Shrike (vr), Water Pipit (l), Dipper (l), Nightingale (a), Northern Wheatear (a), Blue Rock Thrush (r), Whinchat (l), Stonechat (l), Goldcrest (a), Firecrest (l), Short-toed Treecreeper (a), Eurasian Treecreeper (l), Raven (a), Golden Oriole (r), Serin (l), Sis-

Reptiles and amphibians

Reptiles and amphibians can be seen along all routes. Generally speaking, the dry, warm and sheltered areas (routes 2, 4, 6, 10, 11, 14, 16, 17, 18, 19) support the most species. Any river is suitable for finding Viperine Snakes.

Reptiles

Any walk through a dry scrub or forest will reveal a few reptiles. The vast majority of them will be either Wall Lizards (the small ones) or Green Lizards (the larger, bright green ones). However, throughout the Cévennes, there are many more species of lizards and other reptiles, many of which are quite abundant, but live a secretive life.

This is especially applicable to the ten species of snakes, of which six occur in good numbers, that can be found in the Cévennes either throughout or in certain locations in the region. Hence the abundance of Short-toed Eagles – the only European representative of a mainly African group of raptors, the snake-eagles *(Circaetus)*. But only one snake is regularly encountered; spotting the others requires either luck or a search specifically for them. The common one is the Viperine Snake which, despite its intimidating name, is a harmless water snake of rivers and ponds. A small snake that zigzags across the river is usually a Viperine Snake, although Grass Snakes – equally harmless – are aquatic as well. The Grass Snake, a familiar water snake in most European wetlands, is uncommon in the Cévennes.

The Western Whip Snake is a large snake of light forests. We stumbled on these two mating snakes on a minor road on Causse Larzac.

The other snakes that are at least locally abundant (if rarely encountered) in the Cévennes are Western Whip Snake, Aesculapian Snake, Smooth Snake, Montpellier Snake, Adder and Asp Viper. Of these, only the Adder and Asp Viper have a venomous bite and are therefore considered dangerous (the Montpellier Snake does have a weak venom, but this is injected by teeth at the base of the jaw and is not used to strike in defence). Both Adder and Asp Viper are readily distinguished from all the other snakes by the combination of a heavy head, bold markings on the back and the vertical pupil with the overhanging scale that give them their mean and frowning look. Being the most cold-adapted snake of Europe, it is hardly surprising that within the region, Adders occur only on the Mont Lozère. The range of the Asp Viper is almost the reverse of that of the Adder, occurring throughout the Cévennes except on the high slopes of the Lozère. Asp Vipers occur in all sorts of rocky, shrubby terrain and light woodlands, a habitat it shares with the secretive Smooth Snake and the beautiful, large yellow-green coloured Western Whip Snake. Typically, Western Whip Snakes seek out warm,

south-exposed open woodlands of Downy Oak and well-developed scrub. It loves basking in the sun on old stone walls, but avoids the true heat of the Mediterranean habitats. Curiously, it seems to have a strong preference for calcareous soils, in contrast to the Aesculapian Snake, which occurs primarily in schist landscapes. Western Whip Snakes are common in the gorges and on the warmer Causses. The best chance of spotting one is by scanning suitable places where it sunbathes in the morning (old stone walls, dry grasslands, dirt roads, etc.).

Wall Lizards occur in any rocky terrain. The swollen body of this female indicates that more are on the way.

In the warm Mediterranean scrublands, the Western Whip Snake's place is taken by the Montpellier Snake, with its large staring eyes and its almost unmarked grey skin. The Montpellier Snake is one of the largest of the European snakes and often climbs in bushes. Montpellier Snakes share

The beautiful Green Lizard is a common species of scrublands and light woodlands in the valleys. Only the males have –in spring– such a striking blue throat. Nevertheless, the large size and bright green body make Green Lizards impressive animals, regardless of the sex or the season you find it in.

their habitat with two other typically Mediterranean snakes, the Southern Smooth Snake and the Ladder Snake, the latter of which only just reaches the southern Cévennes. Finally, the Aesculapian Snake occurs in woodlands in the river valleys and Chestnut groves. Aesculapian Snakes are, like Montpellier Snakes, good climbers and often effortlessly glide up into the Chestnut trees to hunt eggs and chicks.

Lizards are more easily found than snakes, although there are fewer species present in the area. The most common are the Wall and the Green Lizards. Wall Lizards can be found in any rocky terrain (including old walls and houses). It is the lizard you typically see darting along a wall when you are enjoying a refreshment on a terrace or when you are taking a break en route. Green Lizards are distinctly larger and a little more secretive. They rummage around in scrub, rough grassland and in the leaf litter of forest edges. Both males and females are easily recognisable by their bright green colour. On top of that the males have stunningly bright blue throats. On the southern Causses and locally in the Cevenol valleys an even larger and bulkier Lizard is found. This is the Ocellated Lizard, a western Mediterranean Lizard that reaches its northern limits in the southern Cévennes. It is a rare species here, favouring dry, hot, open terrain. Another truly Mediterranean Lizard of the Cévennes is the Iberian Wall Lizard, a confusing lookalike of the widespread Common Wall Lizard. The Iberian Wall Lizard is locally common in the hot scrub of the eastern Valleys of the Cévennes and on the Causse Méjean.

Up on the cold plateaux of the Mont Lozère, three other lizards occur: Sand and Viviparous Lizards and Slow Worm. The first two, which are the two typical lizards of most of Europe, are restricted to the heights of the Lozère. The Sand Lizard occurs here in a wide variety of habitats, whereas the Viviparous is restricted to open, fresh habitats like heathlands and fens. Finally the legless Slow Worm occurs throughout the Cévennes, but is generally quite rare except on the Mont Lozère, because it favours humid areas with loose soil or moist leaf litter for shelter.

Finally, the European Pond Terrapin occurs in some of the larger rivers just outside the region. There is a large population in the Camargue and individuals are occasionally seen on the edge of the Cévennes.

Amphibians

In the introduction of this section (page 71) we characterised the Cévennes as a region blessed with a very high diversity of plant and animal species, because ranges of the warmth-loving species of the Mediterranean overlap in the Cévennes with those of the cold-adapted species of the north. This is why there are so many reptile species.

Strangely, for amphibians the reverse seems to be the case. Most northern species reach their southern limit just north of the Cévennes, whereas the southern species remain in the lowlands and are not found in the mountains either. The Tree Frog for example reaches south to central France while its southern congener the Stripeless Treefrog reaches only the southern extremities of the region.

The Midwife Toad has freaky, golden eyes.

Only three species of amphibians are widespread throughout the Cévennes – the Common Toad, the Midwife Toad and the Marsh Frog, although the status of the latter species is not quite certain because of possible confusion with the very similar Iberian Water Frog. Some authors say the latter is the naturally occurring species in the region, but Marsh Frogs have been introduced and may now be the dominant species. In addition, Grass Frog occurs on the higher elevations and Graf's Hybrid Frog may be found in damp areas in the southern and eastern foothills. The beautiful black and yellow Fire Salamander is found in mountain forests.

Interestingly, the best habitats to find some of the rarer species of amphibians are not the wettest areas in the central Cévennes, as one would expect from the aquatic amphibians, but the Causses. With their shallow and quickly warming water, the lavognes (see text box on page 52) provide a habitat that is rare further north. Furthermore, there are usually no predatory fish in these isolated ponds to eat the tadpoles. The species to look for here are Midwife and Natterjack Toads, Parsley Frog and Palmate Newt. The Yellow-bellied Toad is thought to be extinct in the Cévennes, but this species' remarkable capability to bury itself deep into the soil to escape the drought makes it to be easily overlooked. Any sightings should be reported to the National Park or at Alepe, the natural history society of the Lozère district.

Reptiles of the Cévennes and Grands Causses

Adder (Mont Lozère), Asp Viper (throughout), Viperine Snake (throughout), Grass Snake (locally common), Aesculapian Snake (locally common in river valleys), Montpellier Snake (uncommon in Mediterranean zones), Southern Smooth Snake (rare in Mediterranean zones), Ladder Snake (rare in Mediterranean zones), Smooth Snake (locally common), Western Whip Snake (locally common in limestone areas), Wall Lizard (very common), Iberian Wall Lizard (locally common), Viviparous Lizard (common on Lozère), Ocellated Lizard (rare on southern Causses), Western Green Lizard (common), Sand Lizard (locally common higher up), Slow Worm (locally common)

Amphibians of the Cévennes and Grands Causses

Grass Frog (common on Lozère), Marsh Frog (common?), Iberian Water Frog (common?), Graf's Hybrid Frog (rare if present), Stripeless Tree Frog (very rare in Mediterranean zones), Natterjack Toad (locally common), Common Toad (common), Parsley Frog (locally common on Causses), Midwife Toad (common), Yellow-bellied Toad (disappeared?), Palmate Newt (locally common), Fire Salamander (locally common in mountain streams)

Insects and other invertebrates

The Cévennes and Grands Causses forms one of the more diverse regions in France in terms of invertebrate species, particularly butterflies. Throughout spring and summer the flowery grasslands of the Causses together with the heathlands and bogs of the Cévennes are alive with hundreds of these colourful insects as they feed on the nectar.

Butterflies

There are close to 150 species of butterflies recorded throughout the Cévennes; a number only exceeded by the Pyrenees and Alpine areas. In this section we describe the more interesting species of the region.

The beautiful Apollo flies on the Mont Lozère and Causse Méjean.

Swallowtails and Apollos

The Swallowtails are particularly eye-catching and beautiful. Both the Scarce Swallowtail and the Common Swallowtail can be found, but rather misleadingly, in the Cévennes the Scarce Swallowtail is common and the Common Swallowtail is scarce.

The large, almost transparently white Apollo and Clouded Apollo can be very tricky to locate. Their distribution is scattered and they are best looked for on the Mont Aigoual and Mont Lozère massifs. The attractive Southern Festoon can be found on scrub and dry (flowery) grasslands during spring, when it flies until the start of June.

Skippers

The Skipper family is well represented with no less than 17 species in the region, including several members of the Pyrgus genus, the 'grizzled' skippers. These butterflies are typically found in grassy areas as the food plants of the caterpillars are either grasses, vetches, mallows, cinquefoils or rockroses. Of the grizzled skippers Oberthur's, Foulquier's and Rosy Grizzled are only found on the schist and granite of the Cévennes while Common, Large and Cinquefoil, together with Olive and Safflower Skipper, are found throughout. A good field guide is recommended – and a camera to take reference photos – since some of these species are tricky to separate.

The flight of the tiny Wood White can be characteristically fluttery, and herewith easily distinguished from the other whites.

Whites and Yellows

Nine species of Whites are present in the region, of which the Black-veined White is arguably the most beautiful. Its virgin-white wings with deep-black lines are conspicuous throughout the region. It occurs primarily in flowery meadows, together with Berger's Clouded Yellow and the other, common species of Whites. The small Wood White is usually picked up by its very slow and somewhat weak flight. It is one of the few species that flies in the woodland interiors. The Provence Orange-tip, a Mediterranean species, can be found in the south around the Causse Blandas and the valley of the Vis. Two other, quite similar species are Bath White and Western Dappled White. Both can be found on rocky, flowery hillsides throughout the area. Look out for the olive-green underside hindwing.

Duke of Burgundy Fritillary and Nettle-tree Butterfly

Although the name of Duke of Burgundy Fritillary suggests that it belongs to the fritillary family, it is actually more closely related to the hairstreaks, coppers and blues, and is the only member of its family in Europe. Primroses form the food plants of their caterpillars, so with the many Cowslips (a kind of primrose) across the region it is not surprising to find this butterfly to be fairly common in May and June.

The Nettle-tree Butterfly too, is the only representative of its family in Europe, but unlike the Duke of Burgundy Fritillary, it is impossible to confuse it with anything else. The palps ('mouthparts') are very long and are extended forwards, giving the impression that it has a snout, which is unusual for a butterfly. Like its host plant the Nettle-tree, the butterfly has a distinct Mediterranean distribution. Numbers vary from year to year but it is usually not uncommon in the south of the region.

Hairstreaks

The most common member of the Hairstreak family is the Green Hairstreak, which in spring can be seen in a variety of habitats – grassland, heath, woodland rides and scrubland. Both Spanish Purple and False Ilex Hairstreaks are on the limit of their ranges in the Cévennes and both tend to spend most of their time in the upper branches of oak trees, making them quite difficult to see. Conversely the Blue-spot, White-letter, Sloe and Ilex Hairstreaks can be found in a variety of habitats – grassland, heath, woodland edges and scrubland where they are often attracted to the flowers of Buckthorn and Brambles.

Coppers

The Scarce Copper is an absolute jewel due to its almost iridescent orange-red upper-side, edged with black. It frequents areas around small streams and is a common sight in early summer in the meadows higher up in the Cévennes as well as on the Aigoual and Lozère. You might find the rare Purple-edged Copper here as well, as it prefers damp fields and boggy places. The Sooty and Purple-shot Coppers are both best looked for on grassy slopes and hillsides and are more easily seen on the heaths of the Cévennes rather than the Causses.

You'll find Scarce Copper to be anything but scarce on the heathlands of the upper slopes of the Aigoual.

122

The sky-blue colour of the male Adonis Blue (top) is so vibrant that you can recognise it even when passing in flight. The vittatus-subspecies Furry Blue (bottom) is endemic to the Cévennes.

Blues

The Blues (some of which are brown) are often considered to be the most difficult group to identify, but with a combination of dogged perseverance and a detailed attention to subtle distinctions, all is not lost for the careful observer. Beside physical characteristics of the butterfly (such as their tone of blue or brown), its choice of habitat, flight times and host plants can indicate its identity and so reduce the struggle to arrive at an accurate identification. Certainly it helps to be conversant with basic botany!

The Cross-leaf Gentian is the food plant of the Mountain Alcon Blue and you should be able to locate it around any patch of these tall flowers in June and July. Likewise, Long-tailed Blues are frequently seen feeding on Bladder Senna and Everlasting Pea.

Other species are restricted to certain habitat types. Chequered Blue is found only on dry, stony hillsides and Amanda's and Provençal Short-tailed Blues are found on flowery meadows. There are two species typical of limestone hillsides and grasslands, but they represent the two extremes of the bluish colour pallet. The Chalkhill Blue has a pale greyish-blue colour, whereas the more widespread Adonis Blue sports an iridescent deep blue colour.

Although the combination of habitat, flight time, food plant and general appearance of the butterfly may sometimes suffice to identify a Blue, it is quite often not enough, but it does narrow the options down to two or three look-alikes that need to be examined more carefully. Two very similar species for example, are the Black-eyed Blue, which only occurs in the very south of the region, and the Green-underside Blue that is found throughout. Both species can be found around flowery grasslands. Another confusing pair are Osiris Blue and Little Blue. The first has a blue upperside that helps with identification and is usually found near Sainfoins.

It is on the slopes of Lozère and the Western Cévennes that the Western Furry Blue, an endemic subspecies to this part of the Central Massif, can be found. But don't jump to conclusions as its flight time coincides with that of Damon Blue and both species share a white streak on a brownish underside. Large and Baton Blues are found where there is an abundance of wild Thyme and plenty of ant's nests. Several species of Blues have an interesting symbiotic relationship with ants in their larval stage, in which the ants are tricked into fostering the caterpillars.

In general, rough, flowery meadows draw most blues; Large, Baton, Escher's and Chapman's Blues all of which are at home on the Causses or in the Cévennes. With some luck, you might find Meleager's Blue, the female of which in particular has a heavily scalloped hind-wing.

Mont Aigoual and Mont Lozère host Mountain Argus, that resembles Brown Argus, which is frequent at lower altitudes.

Browns
The Brown butterflies can also cause a few headaches. Like the Blues, there is an abundance of them. Characteristic for Browns is not their colour, but the fact that they always have one or more eye-spots on the wings. Most species found here lay their eggs on grasses, mainly on fescues and bromes, which are present throughout the region. This means that the larval host plant is, in this case, not a very helpful tool for identification.

With one exception the Browns are not the most conspicuous of butterflies; that exception is the Great Banded Grayling, a large butterfly with a prominent white band on the blackish upperwings. It is quite common. Apart from the Great Banded, various other species of Graylings live in the Cévennes.

Probably the most common of all the 'Brown' butterflies in the Cévennes is the Small Heath. It is a plain butterfly, with no real distinguishing features, and it can be found in grassy areas anywhere. It occurs together with Dusky, Pearly and Chestnut Heaths, which are more attractive as they have pale patches and a series of white spots (ocelli) on the hind-wing underside.

Completely different in appearance, but belonging to the same family nonetheless, is the conspicuous Marbled White, which is found on any flowery patch, particularly when there are Knapweeds around. After the Small Heath, it is probably the most numerous species in the region. Marbled White can only be confused with Esper's Marbled White, one of those Mediterranean species only found in the southern part of the Grands Causses and Cévennes.

Common Gatekeepers can be found throughout the region, while the Southern Gatekeeper, which flies earlier in the year,

The False Grayling (top) occurs on warm, poor grasslands Females drop their eggs at random in the vegetation, where caterpillars find grasses to feed on.

Dusky Heath (bottom) is a Mediterranean species that reaches the northern border of its distribution in the Cévennes.

occurs only in the southern Cévennes. The Spanish Gatekeeper does not venture much further north than the Causse Larzac and the best place to look for it is on Causse Blandas from May to July.

The Ringlets *(Erebia)* form a large genus of delicate blackish-brown or sooty butterflies, often with a red band. There are many similar species and almost all of them are restricted to mountainous regions. Approximately ten species of Ringlet are found in the mountains of the Cévennes. The Scotch Argus prefers damper grassland and woodland, while the Autumn Ringlet and Piedmont Ringlet draw to the dry and open grasslands. Arran Brown, Large Ringlet, Bright-eyed Ringlet and Mountain Ringlet are three upland species which are mostly seen in light woodland or on heaths and grasslands.

Admirals, Emperors and Tortoiseshells

This large family of butterflies comprises some of the most beautiful, as well as the most familiar, butterflies, such as Peacock, Painted Lady and Comma. All of these, plus some less familiar species, occur in the Cévennes.

The name 'Admiral' is a contraction of the apt word, 'Admirables', by which they were known in earlier times. Apart from Red Admiral, two other "admirables" can be found here. The White Admiral, a northern European species, can be found in deciduous woodlands, often quite high in the canopy, and is known for its habit of gliding for quite long distances. In the south of Europe, the Southern White Admiral gradually replaces this species. Like many species with similar distributions, the ranges overlap in the Cévennes. Southern White Admiral is usually found in light woodlands and scrubby areas, a habitat it shares with the Large Tortoiseshell. The latter hibernates as adult and emerges when weather conditions are favourable, which may be as early as February.

The Purple Emperor feeds on soil minerals and animal dung and is rarely seen on flowers.

The Camberwell Beauty, an incredibly striking butterfly, is another species that – unfortunately – spends most of its time up in the canopy of tall trees.

Both the Purple Emperor and the Lesser Purple Emperor (with their shiny purple wings arguably the most beautiful of all European butterflies) occur throughout the area, but utilise quite different habitats. The Purple Emperor is usually found in mature woodlands and spends most of its time around the tops of tall oaks (it can be really crowded up there). The Lesser purple Emperor prefers light woodland, often close to rivers and can sometimes be seen taking minerals from the damp soil (or animal droppings) by the riverbanks. The beautiful but rare Two-tailed Pasha is associated with the Strawberry Tree on which it lays its eggs. It is restricted to the warm southern parts of the Cévennes.

Fritillaries

Another complex family, the fritillaries can roughly be ordered in four groups; The *Argynnis* group, *Boloria* group, *Melitaea* group and – most colourful – the *Euphydryas* group.

Fritillaries often require a good view of the details of the underside of the wing to be able to differentiate between species. Like in many mountainous areas, the species diversity in the Cévennes is impressive and forms a real challenge to the identification skills of naturalists. Fortunately, the two largest species, the Silver-washed Fritillary and the Cardinal, are easier to identify. The Silver-washed Fritillary is by far the most common and can be seen in woodland clearings and rides (often on large umbellifers and purple-flowered plants) and on fresh flowery meadows. The Cardinal, which is slightly larger even than the Silver-washed, is only found in the southern part of the Cévennes and Causses.

The Queen Of Spain Fritillary, with remarkable large silvery spots on the underwing, has the longest flight season – March to October – of any fritillary species found in this region. High Brown Fritillary can be found in woodland clearings, while the Niobe Fritillary, which it closely resembles, flies over open grassland and alpine meadows. Black veins distinguish Niobe from High Brown Fritillary.

The Silver-washed Fritillary can often be found feeding on brambles.

Perhaps the most difficult to identify are the fritillaries of the Melitaea group. The main characteristics of this group are found in the underwing patterns. Some look-alikes, such as the Glanville and Knapweed Fritillary both frequent warm flowery meadows and hillsides. Also Heath, False Heath, Meadow and Provençal Fritillaries share a similar habitat and can be common on flowery meadows and woodland edges.

In the Boloria group, Pearl-bordered and Small Pearl-bordered Fritillaries have quite similar underwing patterns, but, apart from the size difference, prefer slightly different habitats. The Small Pearl-bordered is found in damp, flowery forest clearings and moist heathlands, whilst the Pearl-bordered prefers woodland clearings. In flowery meadows above 800 metres you can come across a more alpine species, Titania's Fritillary, which lacks the 'pearl' edging to the underwing. This species can be found together with Marbled and Lesser

Marbled Fritillaries. The latter is also common on the hillsides and in the valleys and gorges.

Although the name suggests it should be expected only in wetlands, the Marsh Fritillary also occurs on dry open terrain like heath and grassland. It has a unique colouration both on the upper and underwing, making it one of the more attractive fritillaries. This Europe-wide endangered species is fortunately still quite abundant in the Cévennes and Grands Causses.

The Cévennes harbours approximately 150 species of butterflies (as a comparison, the UK has 60) and there are few places in Europe with such diversity. This makes it a challenging area if you are a starting butterfly watcher, but for both the experienced enthusiast and for those who just like to see an abundance of colourful butterflies, the Cévennes is one of the finer locations in Europe.

Characteristic butterflies of the Cévennes

Alpine / Mountain species Apollo (Parnassius apollo), Clouded Apollo (Parnassius mnemosyne), Berger's Clouded Yellow (Colias alfacariensis), Mountain Alcon Blue (Maculinea alcon), Mountain Argus (Plebeius artaxerxes), Titania's Fritillary (Boloria titania), Piedmont Ringlet (Erebia meolans), Mountain ringlet (E. epiphron), Bright-eyed Ringlet (E. oeme)

Cévennes' specialities Cinquefoil skipper (Pyrgus cirsii), Rosy grizzled skipper (P. onopordi), Foulquier's grizzled skipper (P. foulquieri), Cleopatra (Gonepteryx Cleopatra), Provence Orange-tip (Anthocharis euphenoides), Osiris Blue (Cupido osiris), Black-eyed Blue (Glaucopsyche melanops), Western Furry Blue (Polyommatus dolus), Damon blue (P. damon), Nettle-tree Butterfly (Libythea celtis), Two-tailed pasha (Charaxes jasius), Cardinal (Argynnis pandora), Provincal fritillary (Melitaea deione), Esper's Marbled White (Melanargia russiae), Autumn Ringlet (Erebia neoridas), Southern Gatekeeper (Pyronia Cecilia), Spanish Gatekeeper (P. bathseba), Dusky Heath (Coenonympha dorus)

Conspicuous species Scarce Swallowtail (Iphiclides podalirius), Black-veined White (Aporia crataegi), Bath White (Pontia daplidice), Green Hairstreak (Callophrys rubi), Common blue (Polyommatus icarus), Adonis blue (Polyommatus bellargus), Southern White Admiral (Limenitis reducta), Queen of Spain fritillary (Issoria lathonia), Glanville fritillary (Melitaea cinxia), Knapweed fritillary (Melitaea phoebe), Meadow fritillary (Melitaea parthenoides), Great Banded Grayling (Brinthesia circe)

Dragonflies

From late April to late September, dragonflies grace the numerous streams and brooks in the Cévennes, the drinking pools or Lavognes on the Causse and the bogs on the Mont Lozère. Each of these waters forms a different environment each supporting a different set of dragonfly species.

The Lavognes typically are shallow water bodies that warm up quickly. Few water plants grow in them. This makes Lavognes a kind of pioneer habitat, attracting species that have specialised in surviving in these types of waters. The most typical species here is the conspicuous Broad-bodied Chaser – a widespread species of central Europe and the UK. This beautiful dragonfly occurs together with other familiar species such as, amongst others, Black-tailed Skimmer, Blue Hawker and Common Darter. Small Bluetail and Variable and Azure Bluets are common damselflies. There are two species at the lavognes more typical of southern European regions: the Southern Skimmer and – more rarely – the diminutive Dainty Bluet.

The peaty wetlands on the Lozère support quite a different population of dragonflies. Not surprisingly, most of them are at home in the bogs and mires of northern Europe and the Alpine regions. Apart from the Large Red Damselflies which are common on the mountain, you may find Moorland Hawker, Downy Emerald, Black and Yellow-winged Darters, Four-spotted Chaser and even the Small Whiteface, a species that is usually found further north. For the south of France, most of these species are as much a rarity as the habitat in which they occur.

The Common Goldenring is one of the largest dragonfly species of Europe and can be found at small streams where males defend their territory against other males.

Blue Featherlegs is, together with its lighter cousin the White Featherlegs, a frequent sight along the rivers and streams in the region.

For the visitor from northern Europe, the most interesting places in terms of dragonflies are the warm, Mediterranean streams in the Cévenol valleys and the Gorges of the Tarn, Jonte and Dourbies. During the summer months, dragonflies abound along the streams here. Even if you are not particularly interested in identifying them to species level, the metallic-purple, fairy-like Demoiselles are enchanting. It is a joy to simply watch the way they dance over the water and land to rest on an overhanging branch. Two species occur together: the Western and the Copper Demoiselles. Equally in evidence are the faster and fiercer Small Pincertails, which patrol in great numbers along the rivers and can also be seen basking in the sun on gravel banks. They are joined by the very similar, but rarer, Large Pincertail and Keeled and White-tipped Skimmers. Among the damselflies, the White and Blue Featherlegs can be quite numerous. The massive, yellow Common Goldenring, one of Europe's largest dragonflies, inhabits similar habitats, but is more often seen in the open woodlands and woodland clearings further afield. They are powerful flyers that come there to hunt. Likewise, the slightly smaller Sombre Goldenring, and some of the more common large dragonflies can be found hunting in woodland clearings.

West of the town of Millau, the Tarn river leaves its dramatic gorge landscape and becomes a more tranquil river. Even though this area is just outside the range of this book, it deserves a mention, because of its dragonflies. With quiet corners of tranquil, standing water, the mid-section of the Tarn river forms a special habitat where three very rare species occur.

130

The most conspicuous group of dragonflies is, without doubt, that of the Demoiselles, which fly around by the dozens over Cévennes' rivers. The species on the photo is a Western Demoiselle.

The Pronged Clubtail is a rare dragonfly with an oddly scattered distribution in Europe; it is restricted to the Tarn and Garonne and a few regions in the southwest of the Iberian Peninsula. The large and impressive Splendid Chaser has a similar distribution range. Finally, the Western Spectre occurs alongside the shaded rivers in most of the south-French lowlands.

List of dragonfly species

Southern and local specialities Western Demoiselle (Calopteryx xanthostoma), Copper Demoiselle (Calopteryx haemorrhoidalis), Dainty Bluet (Coenagrion scitulum), White Featherlegs (Platycnemis latipes), Pronged Clubtail (Gomphus graslinii), Sombre Goldenring (Cordulegaster bidentata), Splendid Chaser (Macromia splendens), Western Spectre (Boyeria irene), White-tipped Skimmer (Orthetrum albistylum), Southern Skimmer (Orthetrum brunneum)

Northern species Moorland Hawker (Aeshna juncea), Downy Emerald (Cordulia aenea), Black Darter (Sympetrum danae), Small Whiteface (Leucorrhinia dubia)

Other conspicuous species Robust Spreadwing (Lestes dryas), Large Red Damselfly (Pyrrhosoma nymphula), Blue Emperor (Anax imperator), Broad-bodied Chaser (Libellula depressa), Blue Hawker (Aeshna cyanea), Small Pincertail (Onychogomphus forcipatus), Great Pincertail (Onychogomphus uncatus), Common Goldenring (Cordulegaster boltonii)

Other invertebrates

We have lumped here, at the end of the flora and fauna section under the term 'other invertebrates', what is actually the vast majority of the total of species present in the Cévennes. It encompasses thousands upon thousands of species of mini-beasts; beetles, grasshoppers, bugs, flies, wasps, ants, spiders, centipedes, millipedes and other "creepy-crawlies". Like in any area of the world, most of these species in the Cévennes go unnoticed and are hardly dealt with by field guides, which is why this section is so short. However, in comparison to some other areas, there is a relatively large number of 'other invertebrates' in the Cévennes that are not so inconspicuous, but brightly coloured, big and very beautiful.

Stag Beetles lay their eggs in dead or rotten oak wood. It is a long-living species and the larvae take at least seven years to develop.

One insect that directly attracts anyone's attention is a fast-flying, restless animal. The first time observer usually sees it as a yellow dash that flashes by over sunny flowery grassland. Is it a butterfly? Is it a dragonfly? No, most likely you are seeing an Ascalaphid. This superb beastie with its black body and flashing yellow wings is a relative of the Ant-lions. They spend their days hunting little insects over the grasslands, occasionally resting and sunbathing on a nearby flower. There are several species of Ascalaphids in Europe and all occur in the (sub)-Mediterranean region, although we have never seen as many as in the Cévennes. The most numerous species here is Yellow-winged Ascalaphid* (*Libelloides coccajus*).

Another very typical insect of the Mediterranean is the Cicada. It is well camouflaged, but, due to its phenomenally loud, rasping sound, cannot be missed. Cicadas only occur in the hot lowlands of the eastern and southern Cévennes and in the gorges in the Causse area. Cicadas 'sing', for lack of a better word, all through the day, but seem to be particularly evident in summer afternoons, when everything else falls silent and the heat wears out even the most enthusiastic walker. The animal itself, which looks like an oversized grey fly with a very broad flat head, is very difficult to find. It usually sits on the bark of a tree where it virtually disappears against the background. Even though, by most standards, the Cicada isn't considered very beautiful, it is part of the southern French identity. Particularly in the Provence, 'la Cigale' is a popular name for houses and the animal itself is depicted in ornamental garden pottery.

The eye-catching Ascalaphid looks like a cross between a dragonfly and a butterfly, but is in fact related to the ant-lions.

The flowery fields are home to several other conspicuous insects. The pink flowers of thistles, Meadow Claries and Scabiouses, are more often than not occupied by one or more small, metallic coloured rose chafers, longhorn beetles, grasshoppers or bugs of various different species. In the grassy vegetation alongside the streams, the bright blue Hoplia Beetles catch the eye. They suck plant saps from the juicy riverside vegetation. They are quite abundant in places, appearing like a scatter of little blue mirrors over the vegetation.

The huge Stag Beetles occur in the more mature beech and oak woodlands. They are attracted by the lights of campsites and you can hear their buzzing flight on warm nights, heavy like miniature helicopters.

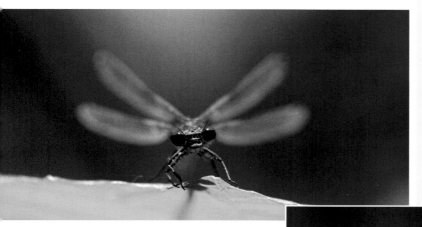

Three conspicuous insects of the Cévennes: the Large Red Damselfly (top), the Blue Hoplia (middle) and Nemophora moths (bottom).

Among the hunting insects are the bright green Praying Mantises. In the hot valleys to the south, Hooded Praying Mantis* *(Empusa pennata)* can be found as well.

Of the long list of spiders, there are four species that catch the eye. The small red tunnel spiders, which we found to be particularly common in the Gorge de la Vis (route 19), the beautiful Orb Web Spider (which occurs in all types of higher grasslands, and the Gravel Wolf Spider (*Arctosa cinerea*; p. 48) and the Wolf Spider. Gravel Wolf Spiders occur solely on gravel banks along the rivers, where they are phenomenally well camouflaged. The large Wolf Spider or Tarantula, is again a Mediterranean species and is found in the scrublands and open, rocky places in the lowlands.

PRACTICAL PART

In this part of the guidebook we recommend routes that are particularly well suited for seeing the many specialities of the area. In addition, from page 215 onwards we provide practical information and observation tips for visitors to the Cévennes and Grands Causses.

This information comes with two caveats. First, remember that things change: roads may change, trails may disappear, books and maps may be sold out, etc. Second, we are not claiming that our listings of things to do are exhaustive. We just offer our experiences by way of helpful suggestions. Our recommendations are not based on some universal standard (if there is one), but on our own experiences, which, we hope, will accord with your own.

The numbered itineraries described in the following pages are shown on the general map in the back flap of this guidebook. The time indications for the routes are based on very leisurely excursions with lots of stops. Whilst the itineraries and maps are straightforward, we still recommend that you also use an official map of the area (see section 'maps' on page 216).

The Cévennes is one of those blessed regions where one really needn't bother to select the sites to visit carefully. Beautiful scenery, seas of wildflowers and a rich wildlife abound throughout the region. The true rambler could just seek out the places that look interesting on the map and start discovering there. There are plenty of minor roads from which you have excellent access to the countryside. In fact, anyone familiar with the Spanish countryside will discover that the Cévennes, and indeed most of France, has a much more extensive network of paved roads, most of them narrow, winding and quiet, which offer plenty of opportunities to halt along the way. Similarly, the network of waymarked trails is, with the exception of the Schist Cévennes, extensive.

The Cévennes provide ample opportunities for walks – from brief strolls in interesting habitats to full day treks.

Therefore we have not tried to provide an exhaustive description of all the trails available, but merely offer our informed suggestions. They are carefully selected to cover the complete range of habitats and landscapes of the Cévennes. They represent the best opportunities to find and observe the most interesting flora and fauna of the region, but they are by no means the only sites of interest.

The first routes (1-6) in this guide are car routes. The car has the advantage of allowing you to visit a large number of habitat types and get a little taste of each of them. The other routes (7-19) are hiking routes. These are really the best way to explore the countryside to the full and to find the typical species that inhabit this superb region. We have described both short and long walks. The shorter ones are selected as to connect to the car routes, allowing you to combine driving and walking to have the best of both worlds.

Each route features a number of icons that indicate the special interests (birds, plants, scenery, ecology etc.) of that particular route. The icons are explained on page 5.

View to the Causse Méjean from Les Bondons (route 10).

Route 1: Touring the northern Cévennes

FULL DAY

Introduction to all major habitats of
the Cévennes, including its flora and fauna.
Particularly great for wildflowers and insects.
Can be combined with walking route 10.

Habitats along this route: forest (p. 34), mountain meadows (p. 41) and broom
scrub (p. 33); Causses (p. 49), Rivers and River Gorges (p. 45).

This route is a perfect first introduction to the nature and countryside of
the Cévennes. With several stops for brief ventures into the countryside,
this car route gives you a taste of all the major landscape types. The trip
departs from the schist landscape that is typical of the eastern Cévennes
and continues through the
granite boulderfields of the
Mont Lozère, to return over
the limestone Causse de
Sauveterre and Gorge du
Tarn.

Departure point Florac

Getting there Leave Florac to the north (direction
Gorge du Tarn and Mende)
and turn right at the round-
about towards Le Pont de
Montvert.

1 The first part of the route follows the Tarn through the typical
Cévennes countryside, characterised by broom scrub, rocky outcrops,
chestnut groves, woodlands and on more level ground, small cultivations.
The Tarn has cut out a V-shaped valley in the schist rock here, very unlike
the limestone region that features the end of this tour, where the Tarn has
created the deep, dramatic abyss of the Gorge du Tarn.

Route 1 skirts the
green Broom-and-
Boulder landscape
of the Lozère.

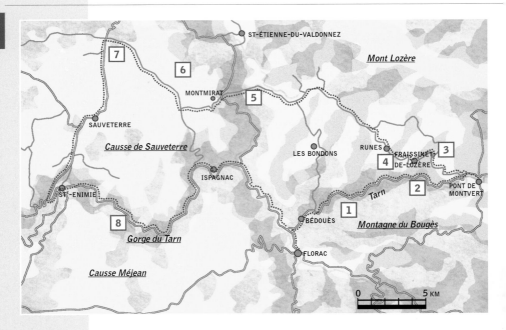

Just beyond the village of Cocurès there are several places to park and where you can follow some of the short trails down to the river. On warm days, on such a descent you should see Wall and Green Lizards, which are very abundant here. The warm, flowery plots are teeming with butterflies and Ascalaphids. In the Tarn, look out for Dippers and Grey Wagtails. In summer, the perfect-blue Hoplia Beetles can be found in the higher vegetation along the river, whilst the shingle along the river is interesting for the rare and large Gravel Wolf Spider* (*Arctosa cinerea*; p. 48).

2 Upon arriving in the charming town of Pont de Montvert – famous for being the place where the Camisard revolts began, see page 59 – the schist landscape has given way to the Granite of the Mont Lozère. Pont de Montvert houses a good information centre on the natural history of the Mont Lozère. Crag Martins dart over the river Tarn, where, again, you might spot Dipper and Grey Wagtail.
Go back in the direction of Florac. Just after you left the town, turn right towards Fraissinet.

3 You now climb higher up and enter the Broom-and-Boulder landscape of the Lozère. The granite bedrock of the Mont Lozère creates a landscape completely different from the wooded valley of the Tarn you traversed earlier. On the Mont Lozère, the meadows are fresh and flowery, with rounded granite boulders scattered about.

Water is conspicuously present in this landscape, creating many small streams and boggy wetlands. The marshes and meadows are particularly interesting for plants, such as Pheasant's-eye Daffodil, Heath Spotted Orchid and Marsh Lousewort providing the botanical interest. In spring, listen for the Ortolan Buntings amongst the more common Yellowhammers.

Continue along the road. Just before arriving at Runes you reach a small car park on the left, signposted 'Cascade de Runes'. Park here for a short, but beautiful, walk to a waterfall that tumbles down from the slopes of the Lozère.

In the pine plantation on the Causse de Sauveterre (point 7) you'll find plenty of Green Wintergreen and Creeping Lady's-tresses (next page).

140

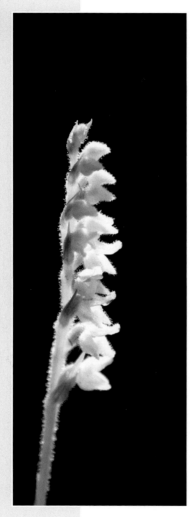

4 The short descent to the waterfall is a good way to see the granite landscape of the Mont Lozère up close. In the stone wall along the path you can find Wall and Green Lizards. In spring, the wet meadows are full with Pheasant's-eye Daffodils. Further down you cross open woodland of Sessile Oak.

After visiting the waterfall, continue on the same road. The landscape becomes more open and cultivated. Right after passing the road towards Les Bondons, there is a parking place on the left hand side, signposted 'menhirs'. From here there is a beautiful roughly two hour walk (route 10), which is described separately on page 180.

Continue along the road towards the Col de Montmirat.

5 After the schist landscape in the valley of the Tarn and the granite landscape around Fraissinet, it is now the limestone soil of the Causse that commands your attention. The transition from granite to limestone is at the Menhir site. You are now, geologically speaking, in Causse landscape, even though you don't enter the Causse de Sauveterre until you have passed the Col de Montmirat a little further ahead. If you stop along the way and look at the soil, you'll find it to be dry and rocky with a thin layer of vegetation which flowers in profusion between May and July. There is more of this ahead!

Pass the Col de Montmirat, which is the northern gateway to the Cévennes, and go straight on to the Causse de Sauveterre. Follow this road for some 3 kilometres until you reach the D31 road. Turn right.

Creeping Lady's-tresses

6 The D31 is a rather busy road so to explore the Causse habitat here it is better to turn off onto one of the minor roads on the right. Here you can pull over and explore the Causse for a bit. It is basically a huge limestone grassland area and a superb place for naturalists. In spring, many orchids (Lizard, Burnt, Man, etc.) are flowering, in addition to the Feather Grass and the endemic Dark-red Pasqueflower* *(Pulsatilla rubra)*, and Hoary and White Rockroses and many other species. There are Ascalaphids and many species of butterflies. These minor roads here could be enough to keep one busy for the remainder of the day!

Continue along the D31 road and turn left towards Chanac (instead of going down towards Balsièges). A little further on, there is a pine woodland where you can stop.

7 Pine plantations are an important element of all the Causses. Even though 'pine plantation' doesn't sound like a particularly exciting place to explore, the older ones like this one are floristically very interesting. Orchids, particularly, abound here with Bird's-nest Orchid, White and Narrow-leaved Helleborines in spring and Red and Dark-red Helleborines in early summer. Creeping Lady's-tresses, a small white orchid, that grows in mossy coniferous woodland, flowers in June and July, just like Green and Round-leaved Wintergreen. The cereal field across the road is also worth exploring for plants with Venus' Looking-glass being abundant.

Go left over the D986, traversing the Causse de Sauveterre from north to south. This part of the Causse de Sauveterre is quite open, the landscape being typical of that of the southern Causses. The road drops down into the Gorge du Tarn at Ste Énimie (spectacular views on the way down). In the village, turn left towards Florac.

8 The road back to Florac runs through the Gorge du Tarn, adding the limestone cliffs to the list of habitats on this route. Despite the superb views, this part of the gorge is not the best and there are few places to stop and explore. However, just before Quézac there is an interesting section with a very shady and moist gorge woodland.

Selected species of this route

Plants: Green Wintergreen, Venus's Looking-glass, Dark-red Pasqueflower*, Yellow Gentian, Pheasant's-eye Daffodil
Birds: Bonelli's Warbler, Crag Martin
Butterflies: Amanda's Blue, Damon Blue, Weaver's Fritillary

!

Some roads are very narrow and steep.

Route 2: The Causse and the gorges

FULL DAY

Great birdwatching; many vultures and steppe birds.
Life and landscape of the open Causes and superb scenery of the Gorges.

Habitats along this route: Mediterranean scrub (p. 30); Mediterranean forest (p. 35); causses (p. 49) , rivers and river rorges (p. 45)

This route is one of the very best car routes to explore the landscape and wildlife of the Grand Causse. It leads you along the most open part of the Causse Méjean, with the most dramatic steppe characteristics. Along the way you pass several of the typical thick-walled hamlets, cross wavy seas of Feather Grass (early summer) and see drifts of orchids and other wild-flowers (spring). This route is also great for butterflies, but particularly it is a good one for birdwatchers. Apart from the many vultures, which you can often see at close range, you can find most of the typical steppe birds of the Causse.

Departure point Meyrueis

Getting there Leave in the direction Florac and, just outside Meyrueis, turn left up onto the Causse, direction Aven Armand.
Upon ascending the plateau you have wonderful views into the Gorge de la Jonte. After about 6 kilometres turn right, direction Costeguison, Nîmes le Vieux. From here on, you are in the steppes.

Just after Costeguison, turn left.

1 This is a good area for birdwatching. Drive slowly in this section to look for Northern Wheatear, Woodlark, Stone Curlew, Short-toed Eagle, Hen Harrier, Montagu's Harrier, Tawny Pipits etc. Black-eared Wheatear is said to occur here, but is very rare, if present at all. The vultures soar over as they will do on the entire length of this trip. Most are Griffons, but occasionally there will be a Black Vulture and sometimes even an Egyptian

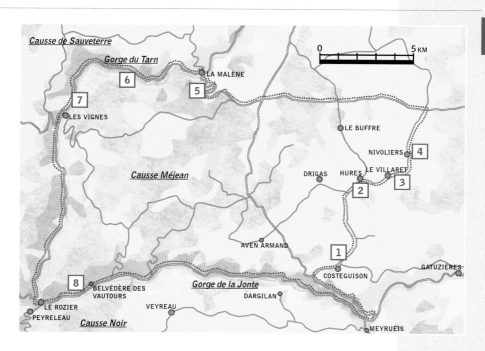

Vulture amongst them. After 4 kilometres, there is a lone tree on the left side of the road. Stop here. This site is particularly good for orchids. Apart from the common Elder-flowered and Burnt Orchids, in May, you can find here the endemic Aymonin's Orchid* *(Ophrys aymoninii)*. Girard's Thrift* *(Armeria girardii)*, endemic to the Grands Causses, grows here as well. In summer, this site, together with the entire south-eastern part of the Méjean is good for watching butterflies.

On the T-junction, turn right.

2 You now come towards the hamlet of Hures. The area around the village is good for observing Stone Curlew. It is not an easy bird to see, but its eerie cur-lééé call sometimes reveals its whereabouts. Just before entering the village there is also a lavogne, just left of the road under a few bushes. On warm days, many birds flock in to drink, sometimes including Rock Sparrows that breed in Hures (especially around the church). Continue towards le Villaret

The steppes of the Causse Méjean look hot and desolate in the summer sun. Despite this appearance, they are among the most diverse areas in terms of wildflowers, birds and butterflies.

3 The rocky area just around le Villaret is one of the best places in the Cévennes for finding Rock Thrush. Check the protruding rocks and the telephone wires. Red-billed Chough is usually present and, in spring, Ortolan Bunting can often be heard as well. Towards your right you can see the Przewalski horses (see text box on page 69).

Le Villaret is one of the abandoned hamlets on the Causse and a lovely place to poke around. There is a little oven house of which the entire dim interior is covered in Maidenhair Spleenwort, which thrives in very shady conditions as long as it is moist enough. Wall Lizards rush around the rocks, but then again, where don't they?

Continue towards Nivoliers, where there is a small bar for refreshments.

4 From Nivoliers southwards, a track (the GR) runs parallel to the hillside. You could follow it on foot for a bit. There are plenty of Causse butterflies and lizards around, while in spring, you can encounter four species of bunting together here: Corn, Ortolan and Cirl Buntings plus Yellowhammer.

From Nivoliers, continue along the road to the D16 main road. At the crossing there is an airstrip for glider planes. This level steppe area can be another place to look for birds typical of the Causses.

You arrive now on the D16, the major – but still minor – road of the Causse Méjean. Turn left and follow it over the Causse. The first part remains an open steppe, but quite soon you enter a large pine plantation that burnt down in 2003. Subsequently, the road descends into a doline. At the crossing, go straight and at the next T-Junction turn right, direction La Malène.

5 The steep descent towards La Malène is spectacular, if not a bit scary on the narrow road. If possible, stop in one of the small car parks and examine the rocks. There are some nice plant species on these slopes, including the endemic Lecoque's Red Valerian* *(Centranthus lecoqii)*. If parking is not an option, don't worry because there are more cliffs further ahead.

In La Malène, cross the Tarn and turn left.

6 This section of the Gorge du Tarn is the most sublime; sheer limestone cliffs tower over the road and in some places there are overhangs which almost push the road into the Tarn. The scenery culminates in the Cirque des Baumes, just north of Les Vignes. At several places there is room to park and examine the cliffs. Apart from the scenery, the flora and wildlife of the gorge are magnificent, even though you'll have to accept that much of it will remain out of sight because of the inaccessible terrain. On the open parts, look for Alpine Swift (usually very high up and more easily seen from the upper gorge edge). On your left side you see some fine beech woods where, beyond any hope of reaching, there are a few localities for Lady's Slipper Orchids. Viewing over the river you can see Kingfisher, Dipper, Crag Martin and Grey Wagtail.

The Cardabelle or Acanthus-leaved Carline Thistle is a typical Causse flower that is intrinsically linked to the life on the plateau (see text box on page 85).

At Les Vignes there is an opportunity to explore the gorge woodland from up close via a tranquil walking trail. In principle you can spend hours walking the gorge here, which wouldn't fit in the schedule set out for this car route, but a short stroll along the track will suffice to get a good glimpse of the gorge woodland including its denizens.

In Les Vignes, cross the Tarn and turn left and pass the campsite (there is an opportunity to swim in the Tarn here). Follow the road for a few hundred metres until you find a suitable parking space and continue on foot.

7 The road soon becomes a track and subsequently a trail following the gorge on a more or less even level. The silence on this track is a bliss after the hectic canoe-and-camper tourist life on the main road. The track leads through open woodland with Downy Oak and Box, alternated with small open patches of abandoned terraces. Look here for orchids and other plants of limestone meadows, butterflies, Ascalaphids, Praying Mantises, Green and Wall Lizards. The mossy undergrowth of the Downy Oak woodlands creates its own cool, moist microclimate where you can find such species as Creeping Lady's-tresses. Birdwise, this little walk has not much to offer other than the usual woodland birds.

Gorge de la Jonte

The Rock Thrush breeds in karst areas on the Causse. The area around the hamlet of le Villaret is a good place to search for them.

Continue down the main road and cross the Tarn again further south at Le Rozier. This village lies at the entrance of the Gorge de la Jonte that separates the Causse Méjean and Causse Noir. The Jonte Gorge is only slightly less spectacular than its larger cousin the Tarn Gorge, but because canoeing isn't possible on the shallow river, there is much less tourism: the Gorge de la Jonte exudes a welcome air of tranquillity and authenticity.

8 Two kilometres beyond Le Rozier, on the left side of the road, the vulture centre is located. This centre was closely involved in the reintroduction of the Griffon Vultures in 1981 and later the Black Vultures in 1992. From the centre you can observe the birds in their colony and watch a wonderful film of the vultures in the Cévennes, including the life cycle of a young bird from birth to its first flight (see text box on page 104).

Continue along the road (there is a nice café 1½ kilometres further ahead). After 17 kilometres you are back in Meyrueis.

Selected species of this route

Plants: Elder-flowered Orchid, Aymonin's Orchid, Yellow-wort, Blue Catananche
Birds: Montagu's Harrier, Rock Sparrow, Stone Curlew, Rock Thrush, Ortolan Bunting, Tawny Pipit, Red-billed Chough, Griffon, Black and Egyptian Vultures
Insects: Red-underwing Skipper, Dingy Skipper, Chalkhill Blue, Meadow Fritillary, Great Sooty Satyr, Apollo (very rare), European Owlfly (Ascalaphid).

148

Route 3: The chestnut groves of the eastern Cévennes

!

Some roads are very narrow and steep.

FULL DAY

Beautiful drive over tranquil, narrow roads in the Schist Cévennes.
An overview of most important habitats of the Cévennes.
Can be combined with walking routes 12 and 13.

Habitats along this route: Mediterranean scrub (p. 30); forest (p. 34), heathland (p. 32); causses (p. 49) , rivers and river gorges (p. 45).

This car trip in the central Cévennes is rather short, but is, with several options to stop and walk, just as complete as the others. It leads through both limestone and schist habitats thus giving a good overview of the flora and fauna of the region, in addition to views of a splendid landscape. The focal point of the route is on the schist Cévennes with its extensive and ancient chestnut groves and its scenic mountain heaths. This route is particularly interesting for its historic landscape, its chestnut groves and the many species of wildflowers and insects.

The pretty Purple-shot Copper is a typical butterfly of the Cévennes' heathlands that you pass on this route. It flies in July.

Departure point Florac

Getting there Leave south in the direction Vébron following the Tarnon River.

1 The route starts in the gorge of the Tarnon, right on the border between schist and limestone soil. If you have an interest in geology, this is a fascinating stretch. The limestone of the Causses overlies the older schist rock, and the road runs roughly on the contact zone between the two (see geology section on page 22).

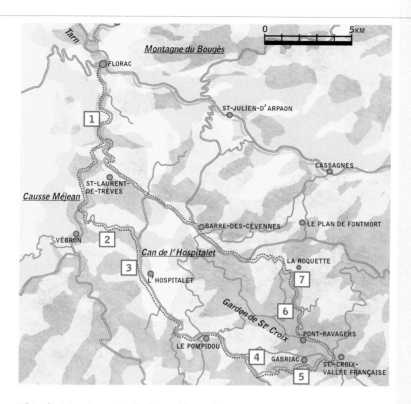

This division is particularly visible at the small mine that you pass along the way.

Ignore the major road towards Barre des Cévennes, but instead go straight ahead and turn left at kilometre 18 on a minor road in direction of Barre des Cévennes and Le Pompidou.

2 This small and quiet road winds up the Can de l'Hospitalet, a small Causse that is part of the National Park. The Can is different from the Grands Causses for having a mixture of both schist and limestone rock. Along this road, the bedrock changes from the former to the latter. About half way up, it pays to stop and stroll along the road for a moment, because in spring, there are several species of orchids (including the fairly rare Woodcock Orchid) and broomrapes in flower.

On top of the Can, turn right towards l'Hospitalet. You are now on the 'Corniche des Cévennes'.

3 The Can de l'Hospitalet, is a limestone plateaux with typical Causse vegetation of shrubs and steppe grasslands. There are very few places to stop here, but you can park at the hamlet of l'Hospitalet and walk route 12. This moderately difficult walk takes about 2 hours to complete and gives a good introduction to the nature of the limestone plateaux. The walk departs from the bar at l'Hospitalet.

Continuing along the Corniche, you traverse the Can and descend its eastern side. There is a viewing point here with a beautiful view over the ridges of the eastern Cévennes.
Upon descending, the rock changes suddenly again from limestone (with, in spring, Lizard and Pyramidal Orchids in the road verges) to schist.
The road continues following the ridge of the mountains. Three kilometres after the village of Le Pompidou, in an old bend of the road, there is a large viewpoint. Park here.

4 Even though this spot is, arguably, a more popular place than the view actually deserves, for the naturalist this is a very nice area to explore. Amongst the protruding rocks you will find pretty much the complete range of schist rock flowers. Following the little trail over the ridge down, you cross a typical heath of the Cévennes, with Bell Heather flowering in profusion in spring and early summer, followed by Heather (Ling) in mid-summer. The endemic Plantain *Plantago holosteum* occurs here, together with Crested Knapweed* *(Centaurea pectinata)* and Lamb's succory. There are many butterflies around, because they use the exposed rocks for "hill-topping" (a typical butterfly behaviour in which butterflies seek out prominent high places as a meeting point for the sexes). Banded Graylings are very common, but also Scarce and Purple-shot Coppers, Cleopatra,

Large Skipper and Marbled White are regular visitors. We found Subalpine Warbler in the bushes below the viewpoint while Honey Buzzards soar over in search of bee's nests.

Continue along the Corniche for 1.5 kilometres before turning left on a minor road towards Falguiere and Fabriac.

5 The road zigzags downhill through the chestnut groves. Some trees are old, dead or dying, but there are also newly planted groves. This is one of the places where the chestnuts have been replanted in order to preserve the traditional landscape of the Cévenol valleys and ridges (p. 68). There are some small fields and orchards around the hamlets. Towards the bottom of the valley, more and more Holm Oaks appear on the side of the road, and the vegetation gradually takes on the character of a Mediterranean woodland. This vegetation type will dominate the south-facing slope further on.

When arriving in St Croix, take the second bridge over the river and turn left. Continue along this road until you reach the hamlet of Pont Ravagers. At the end of it, just before the bridge, turn right, signposted les Roquettes.

6 This minor road, penetrating into one of the more remote valleys of the Cévennes, is arguably the most scenic section of this route. The first part leads through a Mediterranean forest / scrubland with Holm Oak, Tree and Bell Heather being dominant. Age-old Chestnuts grow here and there, gradually becoming more abundant as you climb higher. On damper spots, Bracken Fern covers entire areas. This vegetation is not very rich in species, but certainly attractive.

7 At the farmstead of les Roquettes a short round walk (route 13) has been laid out and takes about 30-45 minutes to complete. It leads you through a very

Sweet Chestnut

old Chestnut Grove and past the old farmstead, which gives a good over-view of the rural life in the Cévennes as it has been lived for centuries.

Continue along the little road through some beautiful chestnut groves. Do not take the turn towards Mas Valat and Le Ranc, but zigzag your way up on the mountain. Almost on the top, you come into a large open heathland area. It pays to stop along the road and wander into it a little to observe this habitat from up close.

The road continues but for some obscure reason turns into a pot-holed dirt track; a stretch that is no longer than twenty metres, before reaching the well-maintained D 13 towards Barre des Cévennes. Turn left here.

Continue to Barre des Cévennes (which lies again on the Can de l'Hospitalet, in limestone soil) and from Barre, follow the signs back to Florac.

View over the Schist Cévennes (point 4 on map).

Selected species of this route

Plants: Woodcock Orchid, Lamb's Succory, Crested Knapweed*, Pale Toadflax, Daisy-leaved Toadflax, Rock Chamomile*
Birds: Honey Buzzard, Green Woodpecker, Red-legged Partridge, Subalpine Warbler
Butterflies: Purple-shot Copper, Large and Red-underwing Skippers, Chalkhill Blue

Route 4: Causse du Larzac

UP TO A FULL DAY

Outstretched and quiet causse landscape with wealth of butterflies and wildflowers.
Interesting historical architecture and templar towns.
Can be combined with walking route 18.

Habitats along this route: downy oak forest (p. 35); Causses and karst (p. 49).

This beautiful car route uses minor roads to tour a section of the immense southern limestone plateau, the Causse du Larzac. It leads through deserted steppes, flowery pastures and odd rock formations, past fields, sleepy hamlets and open pinewoods. It also visits two of the five beautifully restored medieval Templar villages of the Causse. Along the way, there are ample opportunities to stop and stroll around to search for butterflies and wildflowers or just to take in the landscape. As can be expected on a trip over the Causse in spring, there are plenty of orchids.

Departure point Nant (Alternative departure point is Millau; start reading at point 5).

Getting there From Nant, head into the direction La Cavalerie / A9. The road climbs up the Causse du Larzac. Once on the plateaux, take the second small road to the right, signposted 'Gite de Montredon'.

The Woolly Thistle flowers in the height of summer and owes its name to the hairy flower heads.

1 The stretch towards the Hamlet of Montredon leads through the typical Causse landscape, shrubby steppe with rocky soil and woodlands of small pines. Unfortunately, much of it is now fenced off, but there is still a good place to look around at the first dirt track departing to the right. Here you can find a variety of

Chalk-hill Blues fly
well into autumn.

Causse plants and butterflies and
search for Woodlarks, Red-Backed
Shrikes, Tree Pipits, Crested Tits,
Subalpine and Bonelli's Warblers
and Short-toed Eagles. This is
also a first good spot for orchids,
including the rare Yellow Bee Orchid.

At the next crossing, go straight on.

2 After 800 metres on the left hand side, there is a Lavogne. With its concrete basin it is hardly traditional, but it is worth checking for birds that come down to drink. Linnet, Cirl Bunting, Crested Tit and Woodlark are regular visitors at this site. Moreover, this Lavogne is a good spot to watch Nightjars in the evening. Early June is the best season and dusk (around 10 o'clock in June) the best time.

Go back to the crossing and turn right. The road leads through a quiet countryside with pine woodlands. On your left hand side is a military terrain that is reputed to hold the last pairs of Little Bustards of the Causses, but it is likely that they have now disappeared.
After you leave the pine woodlands, take the minor road to the right, signposted 'Atelier du bois tourné' and 'Les Baumes'.

3 The Hamlet of 'Les Baumes' is a few hundred metres ahead. It is a typical Causse hamlet with a special feature: a cave house (troglodyte; p. 59). The stone façade of the house conceals a cave, where people used to live. It was probably built some five hundred years ago. Rock Thrushes inhabit the rocks around the troglodyte house, so keep your eyes open. The flowery meadow in front of the house is interesting for wildflowers and butterflies, while the rest of Les Baumes might reveal a Red-Backed Shrike or a Hoopoe.

Go back to the main road and turn right to continue the trip.

4 Along the next stretch, the landscape changes into an open steppe with scattered shrubs. Keep an eye open for shrikes – mostly Red-backed Shrikes – but both of the larger grey shrikes occur as well. It is because of areas like this, where both northern Great Grey and the Southern Grey Shrike occur, that it was realised that they were distinct species as they do not interbreed. Short-toed Eagle, Hen and Montagu's Harriers also patrol these parts.
Further ahead on your right hand side is another small hamlet, named Potensac. At beginning of the hamlet, at the side of the road, lies a track that is tunnelled for a length of over 400 metres by box bushes. This remarkable lane has been construed to shelter people against the fierce weather conditions on the Causse.

A little further ahead, the road ends on the N9 towards Millau. The route continues to the left, but some 3 kilometres to the right is a parking place with a beautiful panorama over the Gorge de Dourbie, Millau and the famous Millau viaduct.

5 The N9 leads south towards La Cavalerie and traverses more steppe country. Before the highway was constructed, the N9 was a busy road connecting the Mediterranean plain with Clermont-Ferrand and northern France. Now it is a quiet road with broad verges where you can stop at will. Birds similar to those mentioned under point 4 can be found here.

Turn left towards La Cavalerie and follow the signposts 'St. Eulalie' through the village.

6 St. Eulalie lies in a wide valley with fields and woodlands that has a much softer and friendlier character than the open Causse. The village itself is worth visiting. It is the most important Templar village. It was here that the Templar commander lived. In 1641 the then commander, Jean de Bernuy Villeneuve, became fed up with the constant invasion of his commanderie by the whole congregation of the village that passed through en route to the church. In order to gain some peace, he ordered that the entrance of the church be bricked up and a new door be built in the East end. The altar was then moved to the west end of the church, making it unique as in all other Christian churches the altar is at the eastern end.

Upon leaving the village, follow the signs Lapanouse (here you'll find the departure point to the best orchid site of the whole of the Cévennes; this is described in route 18). Just outside St. Eulalie, turn left towards Viala du Pas de Jaux and Roquefort.
The road winds up the hill through nice Downy Oak woodland. Once up the Causse, turn left towards l'Hospitalet du Larzac and later right on the D77 towards Cornus.

Mass vegetation of Pasqueflowers and Elder-flowered Orchids near La Pezade.

7 The first part of this road is not very interesting, but right after the hamlet of Le Figayrol, there is an interesting stop-n-stroll on your left hand side. A track crosses a cattle grid and leads onto a shrubby Causse, which is good for small songbirds, and plants and butterflies in general.

Continue in the direction of Cornus. On the junction with the D65 turn right (Cornus) and quickly after that, left to Caylar. After driving down the D7, which has little of specific interest, turn right to the hamlet of La Pezade.

8 If you are doing this trip in spring, it is highly recommended that you turn right in La Pezade, onto the D140 minor road towards Canals. After 1 km park on a track on your left. The hillside between this site and La Pezade is teeming with orchids, including Frog, Monkey, Lady, Fly, Green-winged, and Elder-flowered Orchids. Pride of place goes to the rare, endemic Aymonin's Orchid*, several specimens of which can be found here (most grow right on the side of the road a little back towards La Pezade).

Some 150 metres further is a site with the largest population of Pasque-flowers we have ever seen. Tens of thousands grow together here, a magnificent sight! This site is flanked by a track on the right, where you can park. Walk along the track for approximately 200 metres and the pasques will come into view, interspersed with a few hundred yellow Elder-flowered

Causse Landscape near La Couvertoi-rade.

Orchids (we never found the red variety of this species in the Cévennes).
Turn back to la Pezade and cross the highway. Continue towards La Couvertoirade.

9 The stretch towards la Couvertoirade leads through open, unfenced (although this changes rapidly) Causse landscape. Superb karst rocks protrude from the plateau. The surrounding limestone grasslands support, predictably, a great flora and many butterflies and birds. This is one of the few sites for Dartford Warbler in the region; look out for Rock Thrush as well. Along this stretch you can stop and wander onto the Causse at will.

10 After what might be the most beautiful stretch of Causse, la Couvertoirade is probably the most beautifully restored Templar town. However, many other people share our opinion; so you will have to zigzag your way through the crowd, particularly in summer.

Aymonin's Orchid, one of the two orchid species that grow only on the Grands Causses and nowhere else in the world.

From La Couvertoirade, turn right towards Nant. Continue to the T-junction with the D7, turn right for about 300 metres, turn left and continue down into Nant.

Selected species of this route

Plants: Yellow Bee, Frog and Aymonin's Orchid, Golden-drop, Pasqueflower
Birds: Short-toed Eagle, Hen and Montagu's Harrier, Rock Thrush, Red-backed, Great Grey and Southern Grey Shrikes
Butterflies: Cleopatra, Hermit, Pearly Heath, Blue-spot Hairstreak, Chalkhill Blue, Adonis Blue

---

Route 5: To the Mont Aigoual

UP TO A FULL DAY

Diverse route with all major landscape types and altitude levels.
Extensive beech forests and a superb view from
the Mont Aigoual.

Habitats along this route: mountain forest (p. 38), heathland (p. 32) Atlantic
mountain heathland (p. 44) causses (p. 49) , rivers and river gorges (p. 45).

On this route you climb from the warm Mediterranean lowlands up to
the cool summit of the 'watchtower of Mediterranean France': the Mont
Aigoual. On the way up you traverse a beautiful landscape with all major
habitats of the Cévennes, from the orchid-rich limestone grasslands of the
Causse, through the heathlands of the schist rocks, and finally arriving in
the meadows, damp forests and scrub of the granite soil.
Of special interest on this drive are the beech woods of the higher slopes of
the Cévennes and the summit of the Aigoual itself, with its wildflowers, its

The beautiful village
of Cantobre looks as
if transported from
Medieval times.

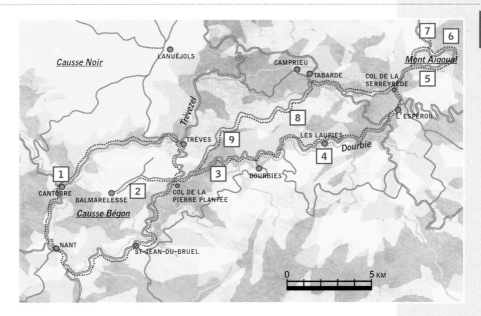

butterflies, its weather station with an exhibition on weather and climate and, of course, its superb view.

Departure point Nant

Getting there Leave Nant in northern direction. After 4.5 kilometres, turn right towards Cantobre and Trèves. You cross the Dourbie River here. Stop at the beginning of Cantobre village, which lies on a limestone escarpment to your left.

1 In our view, of all the picturesque villages of the Cévennes, Cantobre takes the cake. It is a very small, old village that lies on a conspicuous rock right on the point where the Trévezel Gorge ends in the Dourbie Valley. From the cliff's edge you have beautiful views over the valley and of the Crag Martins that fly back and forth from the limestone cliff underneath. The old, cracked house walls are overgrown with rock plants that are typical of the limestone regions, Lizard Orchids are a weed in the small vegetable gardens and there is a Beaver lodge in the Dourbie just underneath the village (visible from the road up to Cantobre). Sheer beauty!

Continue along the road through the Trévezel gorge which is scenic like the more famous gorges (and looking back you have a fairy tale-like view of Cantobre). Any stop along the way will be interesting for the typical plants and butterflies of limestone gorges. Griffon Vultures are likely to fly overhead. In Trèves turn right towards St. Jean de Bruel. At the Col de la Pierre Plantée turn right onto the narrow road (D295) over the Causse Bégon.

2 One can hardly call the Causse Bégon a Causse in the sense of a vast upland plateau like the Causse Méjean and Causse Noir. It is only small and continues in schist mountains just east of the Col de la Pierre Plantée. Nevertheless, Bégon shares the rugged limestone grasslands of its larger brothers, thereby providing an excellent comparison to the schist and granite landscapes that will follow later on. It also has an impressive flora.
There is a wonderful short loop (½ hour walk) here that traverses an area with a good collection of orchids (We counted no less than sixteen species here, including Aymonin's Orchid* and Early Spider Orchid).
One and a half km from the Col, ignore a small road to the left with the name 'Barjac' on a stone. Instead, drive round the next left-hand bend and park at the next right-hand bend where a good track goes off to the left. There is a large grassy area between the road and the track where you can park. Follow this track. After one km another track joins from the right and the orchids will be found in the grassland to your left just a few metres before this junction. Take the track to the right and after about a km turn left onto a tarmac road and follow it through a farmyard (there are dogs here which bark but have never been a problem). Take the next turn to the right and follow this road past another lavogne and you will arrive back at the starting point.

Turn back to the Col de Pierre Plantée and follow the D 151 towards Dourbies and l'Esperou. The limestone of the Causse Bégon soon makes way for schist. At the next crossing, keep following the direction Dourbies. After about 3.5 kms there is a conspicuous viewpoint with a stone picnic table on your right hand side.

3 From this viewpoint you have a great view over the sombre, jagged rock formations above the Dourbie valley. Heathland and small pine trees dominate the landscape. Look out for Short-toed Eagles and Honey Buzzards. There are no obvious trails here, but you can venture on to the heath for a bit and discover some of the typical plant species.

The fresh meadows on the slopes of the Aigoual. In May, they are dotted with thousands of Early-purple Orchids.

Continue towards Dourbies (Robust Marsh Orchids at the village entrance), where the dramatic landscape makes way for the rounded forms that are typical of the granite landscape. Meadows with Yellow Gentian and Piorno Broom* *(Cytisus oromediterraneus)* scrub dominate the views, alternated with orchards and beech woods. In the broom scrub, the flower stalks of the parasitic Greater Broomrape can be conspicuous.
At the beginning of the hamlet of les Laupies there is a car park. Stop here. Walk down the only side road in the hamlet towards the river.

4 Following the river upstream you come to a bridge you can cross. On the other side is a perfect picnic spot. In the little pools in the river are plenty of Viperine Snakes. Demoiselle Damselflies and Small Pincertails are common and on wet spots, butterflies come down to drink and the Dipper patrols the river, in search of caddis fly larvae. In summer you can swim here. A wonderful, peaceful place.
Continuing along the road, you come to the rather ugly ski village of l'Esperou. Follow the signs Mont Aigoual. You then reach the Col de Serreyrède, where there is an information centre and where you can buy

good local produce of the Mont Aigoual region. Continue in the direction Mont Aigoual.

5 Just before reaching the Aigoual, you enter a more or less mature beech-spruce woodland, a rare habitat in the Cévennes, which is largely covered by young woodlands. Opposite the junction of the road towards the Prat Peyrot ski station is a forest track that leads into this woodland. Here and along the road, you might catch a glimpse of a Black Woodpecker, which is a specialist bird of mature woodlands. In spring, hundreds of Wild Tulips flower along this road.

6 A little further on, you reach the final destination, Mont Aigoual, the windy mountain. The unique position of such a high mountain so close to the coast, allows you to see, in good weather, the Mediterranean Sea towards the south, the Causses and the Mont Lozère towards the north and west, and even the outline of the Pyrenees and the Alps. Fog or strong winds, however, are more likely on this mountain, but certainly just as impressive. The Mont Aigoual is often dubbed the deluge capital of southern Europe for its phenomenal autumn and winter showers. There is a record of 950 mm of rain falling on a single night in September! That is nearly a third of the annual rainfall in the wettest spot of Britain! The sudden cooling of warm sea air when it is forced to rise up to the summits of the Aigoual is the reason for the nasty weather. Cold air can't retain the moisture and therefore rains out on the cool mountain slopes. Aigoual is known for its strong, freezing gales as well. The free exhibition on the weather station on top, which is housed in a low, fort-like building, gives a very interesting explanation of meteorology with some extraordinary photographs of extreme weather conditions on this mountain.

Wild Tulips

The subalpine heath-lands on the Mont Aigoual (top) are a rare landscape in the Mediterranean. The Creeping Snap-dragon (bottom) grows between schist and granite rocks along the way up to the Mont Aigoual.

In summer on windless days, when conditions are much more friendly, the area around the top is good for watching 'hill topping' butterflies (a behaviour in which butterflies fly up to the highest point in the landscape to find mates). We even found such Mediterranean species as Nettle-tree Butterfly on the fort walls.

7 North of the actual peak is a ridge, which is treeless. Weather permitting, it is nice to stroll along the GR over the mountain heathlands here. Walk from the parking place on the Mont Aigoual back to the road and follow it left. At the bend to the right, a few hundred metres ahead, go right on the GR that climbs the open hillside. Follow this trail for a bit to enjoy the landscape, the wildflowers and the butterflies of the high altitude mountain heathlands, which are the only ones in the Cévennes away from the Mont Lozère.

From the weather station on the Mont Aigoual, drive back to the Col de Serreyrède and turn in the direction Camprieu. At Tabarde take a sharp left onto the D710 towards Trèves and St. Jean de Bruel.

Some of the most beautiful, moist beech forests grow on the north-facing slopes of the Aigoual Massif.

8 This minor road runs along the same mountain ridge as the one on the way up, but this time you follow the north-facing slope, which is covered by a beech-spruce wood. There are two good sites to stop here. One is at roughly 1.5 km where a sharp bend to the right crosses a small creek. This site is recognisable by a wooden fence. You can park in the side of the road and follow the GR that climbs from here up towards the summit. It's about a 45 minutes (moderately difficult) walk there and back. It leads through some fairly mature spruce and beech wood (some in plantations).

9 The next site is on the road between kilometre 7 and 8, where there is a picnic table at a sharp left hand bend. The area around here has some mature, mossy beechwoods, which are good sites for Black Woodpecker. Walk along the road some 30 metres and you encounter in the verge of the road a small spring with Heath Spotted Orchids and Round-leaved Sundew.

Continue ahead until you reach the D151 again. At Col de la Pierre Plantée, turn left towards St. Jean de Bruel and further back to Nant.

Selected species of this route

Plants: Staehelina, Early-purple, Robust Marsh Orchids, Wild Tulip, Three-leaved Valerian, Sesame Mignonette*, Purple Lettuce
Birds: Crag Martin, Black Woodpecker, Honey Buzzard, Dipper, Tengmalm's Owl
Butterflies: Scarce Copper, Purple-shot Copper, Idas Blue

Route 6: Causse Blandas

2-3 HOURS

Probably the best birdwatching in the whole of the Cévennes.
Empty landscape of scrubs and small villages.
Can be combined with walking route 19.

Habitats along this route: Mediterranean scrub (p. 30); causses (p. 49), cliffs (p. 47).

This short car route, through a beautiful African-like landscape, is especially interesting for finding some of the Mediterranean bird species of the Cévennes. The description below therefore focuses on birds, even though this trip is equally suited for soaking up the landscape or searching for plants and butterflies.

Birdwise, spring and summer bring you species like Montagu's and Hen Harriers, Short-toed Eagle, Hoopoe, various warblers, including the sought-after Orphean Warbler, Ortolan Bunting and many Woodchat Shrikes. The best way to find them is to drive slowly in the early morning or in the evening, looking out for birds on wires and shrubs.

Departure point Alzon

Getting there The tour starts on the D158, just east of Alzon, where you will climb up onto the Causse Blandas.

1 Once on the plateau, start looking for typical Causse species, like Montagu's Harriers, Hoopoes, Little Owls, etc. The first stop is immediately north of a conspicuous rocky escarpment (marked on map IGN 2642ET as 'Belfort'). This is a good site for Rock Thrush.

Continue through Blandas, following the signs for Cirque de Navacelles, and proceed to the viewpoint.

2 From here you have superb views into the valley of the Vis. From the viewpoint, particularly in the early morning, you can see Red-billed

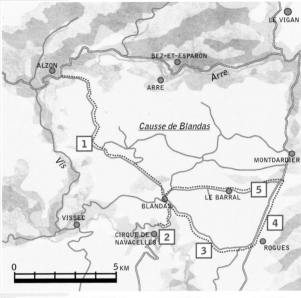

Choughs, Crag Martins, Alpine Swifts and Griffon Vultures flying by. With luck, a Golden Eagle pays a visit as well. In the nearby bushes, look for Melodious and Subalpine Warblers.

Retrace your route to the 'T' junction and turn right, continuing along the D158. Stop at any available place, look and listen.

3 The first section of this quiet road leads you through a very shrubby Causse landscape. Here you will find the typical Causse plants and butterflies, including good numbers of Mediterranean species such as Provence Orange-tip and Esper's Marbled White. Orchid species to be expected at the appropriate times are Lady, Lizard, Green-winged, Monkey, Lesser Butterfly and Early Spider Orchids.

The second half of this road traverses the most open landscape and is best for birds.

The Savannah-like landscape of the Causse Blandas.

Look out for Quail, Red-legged Partridge, Ortolan Bunting, Yellowhammer, Red-backed and Woodchat Shrikes (particularly common), Tawny Pipit, Woodlark, Stonechat and Orphean Warbler. Approximately one kilometre before Rogues there is a huge population of thousands of White Asphodels.

At the end of this road, at a crossroads near a village called Rogues (the English meaning of the word fits the "Western" like landscape very well), turn left onto the D48 towards Montdardier.

4 This is a busier road with less room to pull off the road, but it is still worth looking out as you drive. Rock Thrush is sometimes perched on the electricity wires on the left, with Woodchat and Red-backed Shrikes, Corn Buntings, Cirl Buntings common all around.

About 1 km before Montdardier, turn left onto the D413 towards Le Barral.

5 After about 200 metres, stop at a small turning on the right and look through the gap in the hedge for a Lavogne (water hole), which always attracts birds. It is good for Nightingale, Melodious Warbler, Corn Bunting, Goldfinch and Cirl Bunting, plus many others.

Continuing along the road, scan the bushes and trees in the fields as this is a great place to see Woodchat Shrike in good numbers. Southern Grey Shrikes are also here, albeit much more rare, as are Hoopoes, Turtle Doves and all the previously mentioned species. Watch the skies for raptors, particularly Short-toed Eagles, Black Kites, Buzzards and the occasional vulture.

Birds of Blandas: Woodchat Shrike (top) and Ortolan Bunting (bottom).

At the end of this road, turn left back towards Blandas.

Additional remark This route is easily combined with route 19 that runs through the spectacular Gorge de la Vis.

Selected species of this route

Birds: Golden Eagle, Short-toed Eagle, Hen and Montagu's Harriers, Red-backed, Southern Grey and Woodchat Shrikes, Tawny Pipit, Hoopoe, Rock Thrush, Blue Rock Thrush, Orphean Warbler, Cirl, Rock and Ortolan Buntings

170

! **It can be cold and windy up the mountain**

Route 7: The mires of Mont Lozère

5 HOURS
EASY - MODERATE

Scenic walk through montane beech woods and mountain heathland.
Sub-alpine mires with a, for southern France, rare flora and fauna.

Habitats along this route: mountain forest (p. 38), Atlantic mountain heathlands and bog (p. 41).

This beautiful and easy walk leads you through pleasant cool mountain woodlands up onto the Mont Lozère. Once up on the plateau, you are treated to one of the rarest habitats of the Cévennes, fen mires. Countless springs merge together to form a boggy wetland with mountain heathland on the drier parts.

This walk is interesting for naturalists and for hikers alike.

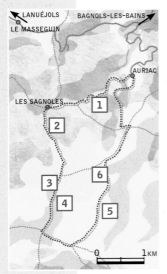

Departure point The hamlet of Auriac on the north slope of the Lozère.

Getting there From Mende, go east towards Bagnols-les-Bains and then turn south onto the D 41. After about 1.5 kilometres, turn left towards the peaceful hamlet of Auriac. There is a small car park at the entrance of the village.

Follow the main track that passes the village on the valley side (not the street into the hamlet). The track is marked by the white-red markings of the GR.

1 The track passes through woodland, alternating with small meadows. There are some spruce plantations, but most interesting here are the beech woods. The fair altitude and the northern exposure provide cool and damp conditions that perfectly suit the plants and

animals of mountainous areas. Many plant species will be familiar to those who have travelled through the temperate regions of Europe. Near the streams there are Fire Salamanders, in the woods you can find Hawfinches and Bullfinches while the road verges are full of wildflowers which attract many butterflies. In particular Arran Browns are common here in summer. In the meadows, the large, metre-high candle-like spikes of the Yellow Gentians attract the attention.

When you arrive at another track, follow it to the right. Parts of the woods have been cleared here, and form another excellent spot for butterflies.

Further on, the track comes to a T-junction at les Sagnoles. Turn left here.

2 The next section leads through an open woodland of old pines and spruces, covered in lichens. The forest floor is a soft bed of Common Feather Grass, mingled with Round-leaved and Heath Bedstraw. The bird-life of these woods is rich, despite the fact that most birds are quite secretive. Among the birds are Crested, Coal, and Marsh Tits, Goldcrests and Firecrests, Eurasian Treecreeper (the rarer of the two treecreepers in the region), Siskin and Crossbill. Ring Ouzel is reputed to breed in low numbers on the edges of the old woods and mountain heaths of the Lozère, and this seems to be a good location.

The mires on the Lozère (the green area in the photo) form a habitat that is very rare in the south of France. The mires host many plants, dragonflies and butterflies that are typical for more northern regions.

Black-veined Whites are common in any grassy roadsides.

Marsh Gentian (opposite page) is one of the beautiful wildflowers of the heathlands of the Lozère.

3 Gradually, the tree cover is getting thinner and a mountain heathland is appearing, with some small boggy streams running through it. Unfortunately, the heath bordering the fields is fenced off, for there are some wonderful plants growing here. However, further on there are no such restrictions.

4 To the left you look out over a large shallow basin with a large flat bog. This route passes it on the way back to Auriac. On your right you see a mountain heathland.

The track splits into two branches that continue parallel to each other. Ahead there is a pass (recognisable by its signs towards Florac and Étang de Barrandon). Some 30 metres before arriving there, take the track that departs to the left. 300 metres ahead you arrive at a crossing upon another pass, marked by a stone cross. Turn left here on the track through the fields, which is lined with barbed wire fences. Stonechats are quite common here, and with some luck you might see a Hen Harrier.

5 The trail runs down through mountain heathland towards the boggy basin you see ahead. Along the way, there are some springs and little streams that run down into the mire. The plants you can find on your walk down are all species you might recognise from trips to the Alps or temperate Europe.

6 The trail passes the bog at its narrowest point where it squeezes between two hills and overflowing water forms a stream that drops towards the valley. A boulder dam keeps the track dry(ish) and it is, particularly in spring and summer, worth to walk over the boulders to explore the other side. Here the fen mire plants are easily spotted from up close, without damaging the fragile vegetation (and getting your feet wet in the process). Look for Marsh Cinquefoil, Heath Spotted Orchid and Bogbean.

!

Do not enter the fen mire itself. It is a very sensitive and rare vegetation that is easily damaged.

Continue over the hill. You now enter some cultivated fields. At the crossing (there is a granite outcrop in the field to your right), turn right. At the next crossing turn left and immediately after that, right, following the signs to Auriac. The descent is now quite steep. There are some good woodlands again just before arriving at Auriac.

Selected species of this route

Plants: Marsh Cinquefoil, Great Burnet, Marsh Gentian, Round-leaved Bedstraw*, Lesser Butterfly Orchid, Cévennes Rock-cress*, Yellow Gentian
Birds: Honey Buzzard, Black Woodpecker, Hen Harrier, Ring Ouzel (rare), Whinchat
Butterflies: Heath Fritillary, Pearly Heath, Arran Brown, Piedmont Ringlets, Chestnut Heath, Marsh Fritillary, Sooty Copper

174

!
It can be cold
and windy up the
mountain

Route 8: The summit of the Mont Lozère

3-4 HOURS (OPTIONAL TO A FULL DAY)
EASY - MODERATE

Open, subalpine heathlands, particularly beautiful in August when the heather is in flower.
The Highest point of the Cévennes, with great views all around.

Habitats along this route: Atlantic mountain heathland (p. 44).

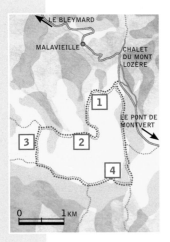

The walk up to the summit of the Mont Lozère, the 'Sommet des Finiels' is an easy walk with only one steepish slope. It leads through the open, rolling heathland that is so typical of the high slope of the Lozère. It is not a very good walk for those wishing to see a great number of plant and animal species, but there is some interesting flora and fauna along the way. In clear weather, the views are superb.

Departure point Le Pont de Montvert.

Getting there Drive towards the Col de Finiels and then another three kilometres. Just after passing the ski lift on your right, park on a small car park on your left. There is a small radio mast here.

Follow the trail in the direction Florac.

1 After passing some rather ugly ski slopes, the landscape becomes interesting. You cross heathlands dominated by Heather and Blueberry. In the small pine trees you can find Tree Pipits, but also the rare Citril Finch, which only occurs in the Alps, the mountain ranges of southern France and those of Spain (with a distinct subspecies in Corsica and Sardinia). In the open heathland, keep an eye open for Skylark, Linnets, Wheatear, Meadow and – rarer – Water Pipits. Viviparous Lizards occur here in the heathland as well. Hen Harriers also breed on the Mont Lozère.

2 The trail takes a left and then a right turn, in which it crosses a few small streams that spring a little further uphill. The streams at the turn are not very interesting, but the one that runs along the enclosure of barbed wire a few metres ahead, is. Botanists may want to walk along this enclosure to find species like Starry Saxifrage, Alpine Willowherb and Round-leaved Sundew. Similar plants grow a little further ahead along the track as well.

Treeless heathlands on the summit of the Lozère.

Roughly a kilometre ahead, there is a sign on the right side of the track, with an arrow pointing both ahead and back. Some 20 metres before this sign, a small track departs to the left, marked by a pile of stones.

3 This moderately steep track, the only somewhat strenuous part of this route, brings you to the summit ahead, which is part of the central ridge of the Lozère. The higher you climb the more open and patchy the heathland becomes. Most plant species disappear further up and make way for a very poor vegetation of only few species that can withstand the freezingly cold conditions: Heather, Bog Bilberry, Bilberry and Alpine Lady's Mantle.

Perennial Sheep's-bit

At this location it is not so much the elevation but the wind that is the greatest enemy of the vegetation. Snow functions like a protective cover of the vegetation, but is blown away on this exposed ridge by the fierce northerlies, leaving the plant parts exposed to the freezing temperatures. In summer, swifts often hunt low over these slopes.

Follow the track to the left.

4 You now ascend the Sommet des Finiels at 1699 metres (so at eye level you are over 1700 metres!). There is not much to see here other than, in clear conditions, the terrific view. To the north you see the Montagne de Goulet and towards the south the Montagnes de Bougès and the Causse Méjean.

Continue along the trail until it comes to a T-junction. Turn left here and walk down (along the artificial menhirs) towards the car park.

Additional remarks Be aware that the weather can be fierce on the top and that there is no shelter at all. Check the weather conditions before starting this trip. If there is some wind in the valley, expect it to be strong at the top; if it is a refreshing breeze, expect it to be cold on top.
This route can easily be extended to twice its original size by continuing along the track after point 2 and turning left after 4 km instead. Use the IGN map 2739 OT to determine your destination and your preferred route.

Selected species of this route

Plants: Alpine Clover, Perrenial Sheep's-bit, Starry Saxifrage, Round-leaved Sundew

Route 9: Mas Camargues

4 HOURS
EASY

Beautiful walk through the open meadows and broom fields of the Mont Lozère.
A wealth of wildflowers, particularly in spring and early summer.

Habitats along this route: mountain forest (p. 38), mountain meadows and heathlands (p. 41), Broom fields (p. 33).

!

It can be cold and windy up the mountain

The landscape around Mas Camargues. In spring the meadows abound with Pheasants-eye Daffodils.

This easy hike over level ground is a good introduction to the landscape of the Mont Lozère. The setting of this beautiful walk reminds one somewhat of an Irish landscape of rolling hills with granite boulders, wide valleys and boggy streams. In spring, the weather can be chilly, but the meadows full of Pheasant's-eye Daffodils and hillsides with abundantly flowering Piorno Broom* *(Cytisus oromediterraneus)* are a feast for the eye. In summer, the cool, green pastures are a pleasant escape from the dry, hot lowlands. In comparison to the other walks on the Mont Lozère in this book, this one offers the most complete range of Lozère habitats.

Departure point Parking spot of Mas Camargues.

PRACTICAL PART

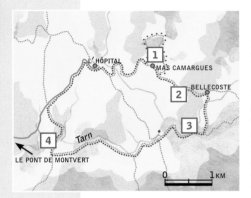

1 The farmstead Mas Camargues dates from 1885 and is an example, although a particularly large one, of the thick-walled buildings of the Mont Lozère. The name is indeed a reference to the famous Rhone delta, the Camargue. The name may suggest a historical connection to Knights of Malta (an order of crusaders) who had their main base in the Camargue. A rival explanation, however, is that it was the summer pasture of the shepherds of the Camargue (see history section).

A small educational trail climbs up the hill above the farmstead. At the time of writing the trail was hardly recognisable, but it is, nevertheless, worth trying.

Follow the track towards Bellecoste and Mas de la Barque.

2 To your right you have a splendid view over the broad valley of the Tarn. In spring the meadows here are full of Pheasant's-eye Daffodils and – less numerous – Aconite-leaved Buttercup (a large, white-flowering buttercup). Both species are typical of Alpine meadows. A few Wild Tulips grow by the side of the track further ahead towards Bellecoste. The dense broom scrub is very poor in species, but Greater Broomrape, as the name implies a parasite on Broom bushes, is quite numerous. Check the wires and bushes for Stonechats, Whinchats, Wheatear, Black Redstart. This is also a good site for the sought-after Citril Finch, which breeds in the hamlet of Bellecoste further ahead.

Just before you arrive at Bellecoste turn right (birdwatchers are advised to continue to look for Citril Finch). The track crosses the fields towards the forest on the other side of the valley. Once in the forest, it ends on a dirt track. Follow this to the right.

3 The trip continues through a rather uninspiring spruce plantation that ends at the Pont du Tarn, a stone bridge over the Tarn river that forms a popular picnic spot. Continue on the south side of the Tarn. The track is more open and traverses a typical mountain heathland. You have beautiful views over the Tarn River, which is broad here and flows through an open landscape; the exact opposite landscape of the narrow gorges the river will squeeze through further downhill. Ahead, the river is dammed, making it an excellent site for anglers. Here you enter another patch of woodland, this time with stunted Beech trees, reflecting the severe weather conditions on the mountain.

The trail ends on a T-junction. Turn right here and walk downhill towards the Tarn.

The upper section of the Tarn.

4 This small area at the banks of the Tarn is the most interesting in terms of wildflowers and butterflies. The moist heathland harbours plants such as Yellow Gentian, Marsh Marigold, Marsh Forget-me-not, Marsh Violet and Heath Spotted Orchid. The stream is home to a number of northern dragonfly species, like Downy Emerald and Four-spotted Chaser. Being sheltered from strong winds, this spot is interesting for butterfly enthusiasts as well.

Cross the Tarn (which consists of a number of small streams here) via the stepping stones and turn right on the other shore. Continue ahead and swing past the hill on your left hand side, until you are on the small road. Follow it back towards l'Hôpital and Mas Camargues.

Selected species of this route

Plants: Pheasant's-eye Daffodil, Wild Tulip, Greater Broomrape, Sesame Mignonette*, Yellow Gentian, Heath Spotted Orchid, Aconite-leaved Buttercup
Birds: Hen Harrier, Dipper, Citril Finch, Crossbill, Water Pipit
Insects: Large Pincertail, Large Red Damselfly, Green Hairstreak, Checkered Skipper

180

Route 10: The Menhirs of les Bondons

!

This trail is very exposed. It can be very windy and the sun can be very fierce.

**2-2½ HOURS
EASY - MODERATE**

Prehistoric menhirs in a steppe landscape.
Walk on the border of the limestone and Granite soil.
Can be combined with car route 1.

Habitats along this route: pine plantation (p. 40), Causses (p. 49), broom fields (p. 33).

This short walk leads you to the largest collection of menhirs (prehistoric standing stones, see page 56) in the Cévennes. The steppe-like setting in which the menhirs are situated gives this area a wonderful, ancient feel (a feel that is much enhanced when doing the walk in the evening). Geologically and botanically, this walk is of interest as well, because it follows the border between the granite landscape of the Mont Lozère and the limestone of the Causse de Sauveterre, thereby displaying the differences in landscape.

Departure point Car park on the D 35, east of the Col de Montmirat, just before the turn towards Les Bondons.

The Menhirs of les Bondons

Le Puech is a strangely shaped limestone hill.

From the car park, follow the trail that leads west along the cereal fields (the track runs parallel to the main road). It leads through some fields and small patches of pinewood with a Causse vegetation that harbours all the typical species, including several orchids, Leuzia, Carduncellus* and Dwarf Thistle. The path ends on a track. Turn left.

1 You encounter the first Menhirs. The track soon enters an area of open pine woodlands. The soil is very bare and rocky, but in spring and early summer, there are many wildflowers. Particularly, Yellow Bird's-nest is numerous here. Birds include Crested Tit, Tree Pipit and Bonelli's Warbler.

The track descends and after a curve it ends on a T-junction. The route turns left, but it is interesting to explore the track to the right for some 200 metres.

2 The sharp and very distinct border between limestone and granite lies in the small valley ahead. Up until now, the route brought you through very dry, limestone soils. In the valley, the grass is higher, there

Wood Pink (top) and Montpellier Pink (bottom) are two common species on this route.

is a trickle of water and Bracken Ferns and Piorno Broom* grow between the conspicuous granite rocks on the hillside ahead. The sound of a small stream confirms the presence of water.

3 Once back on the main route, the track winds through a pleasant rural landscape. The soil here is granite instead of limestone. In clear weather, you have superb views over the hills and in the background, the cliffs of the Causse Méjean.

Just before the picturesque hamlet of Les Combettes, turn left.
The ascent soon brings you into the limestone area. Again you traverse a small area of pinewood, similar to the one earlier. Once out of the pinewood, you enter the most interesting part of the route.

4 Four large menhirs mark the end of the pinewood. To your right you have a superb view over a very open, grassy terrain with two large, bare, rounded hills, les Puechs. These are of a very soft limestone that easily erodes. On top of the hill, you arrive at a Cromlach, a circle of menhirs that is believed to have been used for special religious rituals.

You have now arrived at a small, asphalted road. Turn left to get back to the car park.

Additional remark This route can be walked as a part of car route 1.

Selected species of this route

Plants: Montpellier and Wood Pinks, Leuzia, Dwarf Thistle
Birds: Bonelli's Warbler, Crested Tit, Short-toed Eagle
Butterflies: Damon Blue, Grayling, Large Grizzled Skipper, Hermit, False Grayling

Route 11: Bois de Paiolive

2½ - 4 HOURS
EASY

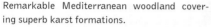

Remarkable Mediterranean woodland covering superb karst formations.
Rich flora and birdlife of Mediterranean scrub.

Habitats along this route: Mediterranean Scrub (p. 30); downy oak forest (p. 35) karst (p. 24), rivers and river rorges (p. 45).

The Bois de Païolive is a superb karst area in the eastern foothills of the Cévennes. It differs from the other karst regions (Nîmes le Vieux, Montpellier le Vieux) in that the chaotic rock formations are overgrown with Mediterranean woodland with wildly winding vines of Black Bryony. This makes the Bois de Païolive an impenetrable jungle with overgrown rock ruins that in places reminds strongly of scenes from the Jungle book. In this chaotic and beautiful terrain three round walks have been set out. The surrounding hills are worth visiting, for it is here rather than on the way-marked trails, that the best flora and fauna is to be seen.

The yellow route follows the edge of the Gorge of the Chazessac. This is a good spot for birdwatching, with species like Alpine Swift, Crag Martin and Black Kite.

Departure point The minor D 252 road cuts through the Bois de Paiolive, just east of Les Vans. Along the D 252 are three parking places. From the second, two round walks depart; the third one departs from the easternmost parking place. On each of the parkings are wooden shields that show the round walks.

Black Kite (top) and Black Bryony (right). The latter is a common climber in the forest of Paiolive.

The Downy Oak forest of Bois de Paiolive has beautiful karst features, including strange rock pillars, caves and cracks (opposite page).

Round walk 1

(Circuit Vert / Green – 'the labyrinth') leads through the most extraordinary piece of karst woodland. The route starts through woodland over level soil with few rock formations peeking out above the canopy, but gradually the rocks dominate the landscape. You find yourself winding through a labyrinth of narrow alleys, ducking rock arches and climbing blocks of stone. Underneath the canopy, the Mediterranean forest shows itself as a warm, moist environment.

Round walk 2 (Circuit Bleu / Blue) is the longest route, leading through a large area of woodland with rocks, and onto a viewpoint over the river gorge. It combines the karst features of the green and the gorges of the yellow route.

Round walk 3 (Circuit Jaune / Yellow) passes over the ridge of a very abrupt gorge of the Chassezac river. The views from here are spectacular and the cliff's edge supports a richer flora and fauna than the interior forest on the other two routes. From the ridge, look for Alpine Swift, Black Kite and Short-toed Eagle. With luck, you may see an Egyptian Vulture.

Additional remarks Because of the exceptional scenery, the Bois de Païolive is a very popular destination for day trips. Avoid weekends and the busiest holiday season. The Bois lies on the calcareous eastern rim of the Cévennes; an area with low rugged ridges separated by level agricultural fields with a superb Mediterranean flora and fauna. The minor roads that pass through the Montagne de la Serre just south and east of the Bois are very

little visited and great for searching for plants and animals of the Mediterranean region.

The nearby village of Banne has a beautiful Medieval centre and a large fort on a hill, from which you can see Alpine Swifts.

Selected species of this route

Birds: Black Kite, Egyptian Vulture, Short-toed Eagle, Alpine Swift, Crag Martin, Sardinian and Subalpine Warblers

---

Route 12: Le Can de l'Hospitalet

2 HOURS
MODERATE

Walk over a small Causse and through a beautiful cliff woodland.

Habitats along this route: forests (p. 34), causses (p. 49).

The trail along the Can de l'Hospitalet.

This short walk crosses the Can de l'Hospitalet, a small limestone plateaux on the edge of the schist region. The Can is geologically extremely diverse and this diversity is reflected in the flora. Apart from beautiful scenery and breathtaking views, the special attraction of this walk is the Beech wood in the gorge and the small calcareous marsh on the far point of this loop. It is also a good introduction to the Causse habitat, although the 'true' Causses have a richer flora.

Departure point Hamlet of l'Hospitalet

Getting there Take the broad track that departs opposite of the bar of l'Hospitalet. It crosses a very meagre 'grassland' that consists largely of tufts of White and Hoary Rockroses. When you arrive at the edge of the woodland ahead, turn right on a smaller track, which goes down the edge of the plateau.

1 The hillside here is rich in plant species, butterflies, ascalaphids, with the odd Praying Mantis turning up occasionally. The view over the cliffs of the Causse Méjean is beautiful and Griffon Vultures are likely to soar by.

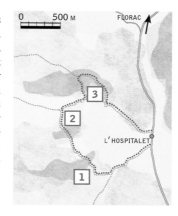

2 After a track from the left has joined the one you are following, you enter an open Beech woodland. A trickle of water emerges from the rock and creates a marshy patch on your right hand side where the trail ends on a broad track. Such calcareous marshes are rare and found exclusively on a few patches under the Causse cliffs where water emerges from the underground. The marshes support a special vegetation with, in this case, amongst others, Bug Orchid and Common Twayblade.

Turn right on the main track.

Hoary Rockrose is among the more prominent plant species of the Can.

3 The track climbs up the Can again, first through a more or less mature Beech woodland and later through pine plantations (with masses of White Helleborine and One-flowered Wintergreen) until you are up the plateau again, from where it is a mere hundred metres to l'Hospitalet.

Additional remark

This short trip can be undertaken as a part of car route 3.

Selected species of this route

Plants: White Helleborine, Bug Orchid, One-flowered Wintergreen, Hoary Rockrose
Insects: Red-underwing Skipper, Dingy Skipper, Chalkhill Blue, Meadow Fritillary, Great Sooty Satyr, Yellow-winged Ascalaphid*

188

Route 13: The Chestnut forests of les Roquettes

½ - 1 HOUR
MODERATE

Ancient chestnut groves and an original Cévenol farm.
Can be walked as a part of car route 3.

Habitats along this route: chestnut grove (p. 37).

Tucked away deep in the heartland of the schist Cévennes, lies Les Roquettes, a group of traditional Cévenol farmsteads. This short walk leads up to one of them, Mas Chaptal, providing a wonderful glance in the Cévennes' rural history and then circles through an ancient chestnut grove back to the departure point. In a magnificent scenery, you get a closer look into the traditional Cévenol way of life.

This walk follows an interpretation trail of the National Park, marked with numbers that correspond to explanations on a brochure that can be freely obtained from a box in the parking place. The following text highlights the most interesting features.

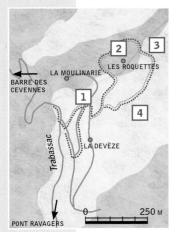

Getting there From Ste.-Croix-Vallée-Française, follow the D13 towards Barre des Cévennes until you reach the hamlet of Pont Ravagers. At the end of the hamlet, just before the bridge, turn right, signposted les Roquettes. Follow the road for about 4 kilometres, where you see a small car park. Park there. A few metres ahead, the trail branches off to the right (see map).

1 The trail runs through a superb old Chestnut grove. The Chestnuts are not harvested anymore, but in former days, they formed the staple diet for the Cévenol people (see page 62; points 4-5 of the interpretation trail).

Once on the tarmac again, follow it taking the next hairpin and then left on another track again (following the signs of the interpretation trail).

189

2 You now walk towards Mas Chaptal, a traditional Cévenol farmstead. The beautiful Mas, drawn up from schist rock with doors and window frames that are, of course, made out of Chestnut wood. Inside you can enjoy a 40 minute film on the traditional life and landscape of the Schist Cévennes. Outside, the trail takes you past some of the characteristic features of a traditional Cévenol farmstead. There is an old threshing-floor that was used for making Chestnut flour (point 8 of the interpretation trail) and bee hives (point 9 of the interpretation trail) for making honey from Chestnut and Heather nectar. At point 11 you encounter a Clède, a building that was meant for drying the chestnuts (see page 64) over a charcoal fire (point 12). Point 13 shows a Black Mulberry (Murier Noir in French), the rarer cousin of the White Mulberry that was used to feed the caterpillars of the Silk Moth during Cévennes' modest 'Golden Age' (see page 63).

The coarse schist walls are interesting for plants, including Creeping Snapdragon and various stonecrops.

Chestnut forest of les Roquettes.

3 The trail then continues and crosses the stream again. In dry periods the stream is a pretty, cool refuge in the baking hills. But in autumn, winter and spring, when the hills draw large deluges from the cloudy skies, the water surges down this gully. The terraces downstream from here have a double function to both slow down the water to prevent erosion and to create more arable land.

4 The trail zigzags through the Chestnut groves. Some of the trees look magnificent in old age, but quite a number are dead or dying. This stage of the woodland is ecologically very valuable because it harbours mushrooms and plants, attracts many insects and birds whilst also providing shelter and breeding sites for birds and mammals. Unfortunately, it also heralds the end of the Chestnut groves.

The trail ends on the track at la Devèze. Follow the track downhill (we found Lizard Orchid in the last hairpin) to the departure point.

Additional remark This circular walk departs from car route 3, but, when undertaken separately, it is recommended to explore the minor road leading up from Les Ravagers.

On this little route you encounter superb old chestnuts. The fact that some of them are dead makes them no less impressive.

Route 14: Walking the Feather Grass steppes

4 HOURS
EASY

Gentle walk through the various landscape types of the Causse, including wide views of the Feather Grass steppes.

A wealth of butterflies and plants, including many orchids and some sought-after birds.

Habitats along this route: pine forest (p. 40), causses (p.49).

!

This trail is very exposed. It can be very windy and the sun can be very fierce.

This route describes a triangle between three charming rural hamlets on the Causse Méjean, which are connected by broad tracks. The hamlets are completely authentic Caussenard structures with massive walls. On your way from hamlet to hamlet you cross the Feather Grass steppes (particularly beautiful in June and July), shrubby, flowery grasslands, lavognes, dolines and open pine woodlands. A good number of the wildflowers (including 13 species of orchids) and butterflies of the Causses can be found here.

You'll encounter Golden-drops frequently along this trail.

Departure point
Le Buffre

This route is described starting from Le Buffre in the north, but can be started at any of the hamlets; it is signposted with yellow bars.

Follow the track that departs from the cross on the village road.

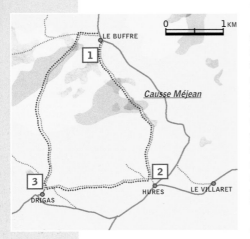

1 Le Buffre (with a few Rock Sparrows in the hamlet, but these are more abundant in the other two hamlets) is situated at the edge of a large and fertile polje (depression in the landscape, see page 24), where cereal is planted. The grassy borders of the track lack the tapestry of wildflowers you will encounter further on. After passing a lavogne (pool) on your right (look for Common, Natterjack and Midwife Toads), the trail gradually climbs a hill with small Pines. As soon as you enter rocky soil, the typical Causse plants appear. Look for Steppe Spurge*, White Flax, Grass-leaved Oxe-eye Daisy* (*Leucanthemum graminifolium*), False Vetch, Mountain Kidney-vetch, Burnet Rose, Spanish Spiny Greenweed* (*Genista hispanica*), Mountain Everlasting, Rouyan's Felty Germander* (*Teucrium rouyanum*) and Golden-drops. In the right season, this is also a good site to look for the endemic Aymonin's Orchid.

Along the entire track, look for vultures and Short-toed Eagles.

Once over the hill the landscape becomes more open and covered by Feather Grass, as you descend to the Hamlet of Hures.

2 Rock Sparrows breed in and around the little church and are quite easily localised by their call (a nasal *zyêêêêêêhh*).

Turn right past the cemetery. Following the track, you climb up another hill, behind which lies a rather ugly quarry. Before reaching the quarry, look back at Hures for a great view of the hamlet amidst the almost Mongolian-looking steppes. Closer at hand, amidst the Feather Grass, there are various species of orchids flowering.

Pass the quarry and follow the track down to Drigas, the third hamlet. This area has many fields, many of which have been abandoned. Look for Stonechat, Linnet, Wheatear and Little Owl and Quail.

In Drigas turn right and follow the yellow signs.

The seas of Feather-grass are a beautiful spectacle from late May to early July.

3 Drigas has a good number of Rock Sparrows as well, so if you failed to see them in Le Buffre or Hures, here is another chance.
Outside of Drigas, you climb another hill. The Feather Grass steppes with several small dolines are particularly rich in orchid species, again including Aymonin's Orchid. Up on the hill, you have great views over the north-eastern part of the Causse and – in clear weather – the Mont Lozère. Look for Dark-red Pasqueflowers* *(Pulsatilla rubra)* here. Your descent leads through young, open Pine plantations (Bonelli's Warbler is very common here).

Once you've crossed the Lavogne of Le Buffre turn right to reach the hamlet again.

Selected species of this route

Plants: Golden-drop, Rouyan's Felty Germander*, Feather Grass, Burnt, Military, Pyramidal, Lizard, Elder-flowered, Fly and Aymonin's Orchids
Birds: Rock Sparrow, Bonelli's Warbler, Woodlark, Short-toed Eagle, Griffon Vulture, Black Vulture, Egyptian Vulture, Tawny Pipit, Tree Pipit, Red-billed Chough
Butterflies: Queen of Spain Fritillary, Berger's Clouded Yellow, Large Grizzled Skipper, Grayling, reputed site for Apollo

---

!

This trail is very exposed. It can be very windy and the sun can be very fierce.

Route 15: Nimes-le-Vieux

**1½ HOURS – EXTENSION TO 4 HOURS
EASY**

Odd karst formations.
Birds and plants of rocky environments.

Habitats along this route: causses and karst (p.49).

Nîmes le Vieux is an extraordinary karst landscape in the south-east corner of the Causse Méjean. The name Nîmes le Vieux – the old Nîmes – was first used as a word play referring to the famous karst of Montpellier le Vieux on the Causse Noir. Both sites feature impressive rock sculptures and pillars separated by narrow alleys, reminding somewhat of the ruins of an ancient city. Apart from the wonderful landscape, Nîmes le Vieux is a very good place to search for the plants, birds and butterflies of dry, rocky areas.

Departure point the hamlet Gally (north-east of Meyrueis).

Leave Gally in the direction of l'Hom via the well-marked trail.

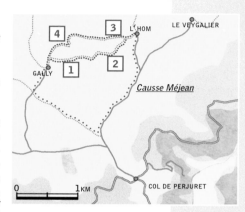

1 The rock pillars of Nîmes le Vieux are located on the south slope of a long east-west hill formation that shelters a large doline valley. In the inhospitable Causse landscape, sites like Nîmes le Vieux were the most suitable sites to build one's farmstead. For the hamlets of Gally, l'Hom and Villeneuve the hill functions as a shelter against the severe winds, and the doline was, and still is, in use for the cultivation of cereals.

The track follows the lower part of the hillside. The first few hundred metres lead over fertile, clayey soil with several large hazel bushes.

The wondrous karst features are reminiscent of a petrified city, hence the name Nîmes le Vieux, the old Nimes.

2 A richer flora is to be found a little further on towards l'Hom, where the soil is thinner and drier. Among the White and Hoary Rock Roses, False Vetch, Mountain Kidney-vetch, Purple Viper's-grass and other familiar Causse plants, we found several orchids, albeit in low numbers: Burnt, Pyramidal, Woodcock, Small Spider and Fly Orchids. For the rarest plant species one has to examine the rock pillars themselves, where you can find Rock Kernera* *(Kernera saxatilis)*, Yellow Whitlow-grass, Alpine Mezereon and Tuberose Valerian. This is also one of the few sites to find the rare endemic Cévennes Saxifrage* (*Saxifraga cebennis*, which grows out of the right 'ear' of the rock that looks like a human head).

Just before l'Hom, turn left and follow the track left that turns uphill and back to Gally.

3 Around l'Hom the birdlife is interesting, with most of the rock dwelling and dry land species present. Look out for Rock Thrush (at the time of writing there was a pair just above l'Hom), Woodlark, Tree Pipit, Little Owl, Red-billed Chough, Cirl Bunting, Rock Bunting and Ortolan Bunting.

Rock Buntings are seen around Nimes le Vieux from time to time. If you don't see them the Ortolan Buntings, Rock Sparrows and Choughs should provide sufficient compensation.

4 The walk back follows the ridge from where you have views of the Causse Méjean to the north. Vultures should be about, with chances of Black and Egyptian Vultures beside the numerous Griffons. Short-toed Eagle and Montagu's Harrier can be found here as well.

Additional remark You can extend this route by continuing from l'Hom to Villeneuve and turn back over the minor road on the south side of the doline.

Selected species of this route

Plants: Kernera, Yellow Whitlow-grass , Cévennes Saxifrage*, Crested Lousewort
Birds: Woodlark, Short-toed Eagle, Griffon Vulture, Black Vulture, Egyptian Vulture, Tawny Pipit, Tree Pipit, Cirl Bunting, Ortolan Bunting, Black Redstart, Rock Thrush, Red-billed Chough.
Butterflies: Queen of Spain Fritillary, Berger's Clouded Yellow, Large Grizzled Skipper, Grayling, reputed site for Apollo.

Route 16: Where the Tarn and Jonte Gorges meet

6 HOURS
MODERATE - STRENUOUS

Very scenic walk with splendid views over the
Gorge du Tarn and Gorge de la Jonte.
Watch vultures passing at eye level.

Habitats on this route: cliffs (p. 47) pine forest (p. 38)

!

**some steep and
difficult passages
along this route**

The village of Le Rozier lies at the junction of the two main gorges, the Gorge du Tarn and that of the Jonte. From here a superb, albeit difficult, trail leads through the cliff forests of the Gorge du Tarn up to the edge of the Causse Méjean, crosses its southwest point and leads back to Le Rozier along the edge of the Jonte gorge. The views along this route are spectacular. Being close to the vulture colonies, you should have wonderful views of vultures soaring through the gorge. Furthermore, this route is perfect to explore all the elements of the cliff forests, particularly its flora.

This corner of the Méjean is the best place for spectacular views of Griffon Vultures that pass through the Jonte Gorge.

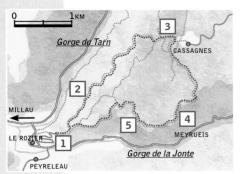

Departure point Le Rozier

Getting there A track leads up to the Rocher de Capluc (the rock with the distinctive cross on top of it) and it is possible to do the first part of it by car. However, there are limited parking possibilities up there, so if it is busy, you may have to park in Le Rozier and walk up through the village in the direction of Capluc.

1 The south-facing exposed cliffs just above le Rozier are dry and sunny and have a Mediterranean character. Look out for Subalpine and Melodious Warblers along the road; some interesting plants include the endemic Silver-lined Broom* *(Argylobium zanonii)*, Lizard and Pyramidal Orchids, Lecoque's Red Valerian* *(Centranthus lecoqii)*. You should see the first vultures soaring overhead.

At the last houses of the Le Rozier, close to the cross of Capluc, the trail splits, the left branch is signposted 'Rocher the Capluc', but you go right instead. After a few metres, turn left in the direction 'St. Pons'.

2 This trail (signposted with yellow markings) is a narrow and quiet one leading along the slope of the Gorge du Tarn. The first part stays more or less level, but after that, it descends to pass underneath some rock formations and leads through a small cave.
This part is scenically splendid as well as botanically rich. There is a mixture of Mediterranean, temperate European and Mountain species and a few endemics (see list at the end of this route). Orchids are numerous here too, with many White, Narrow-leaved, Red, Dark-red, and Broad-leaved Helleborines, Violet and Common Bird's-nest Orchids, Common Spotted, Lizard, Military, Monkey, Pyramidal and Woodcock Orchids and Creeping Lady's-tresses.
After passing through the cave, past the ruins of an old house (with Lecoques Red Valerian* *(Centranthus lecoqii)* growing on the walls), the trail snakes underneath a conspicuous rock pinnacle and zigzags straight up the Causse. It ends at a larger track (you are nearly on the Causse now), which is part of the GR. Turn left here. A little further ahead you come to another track. Turn right.

198

3 You have now arrived on the plateaux, and – after such a magnificent path up to the top – the summit is rather disappointing (the rewarding views are a little further ahead in the Jonte valley). At this point you cross rather dull pine plantations with little of interest, although – it is still the Cévennes – the poor herb layer consists almost solely of orchids. Narrow-leaved and White Helleborines grow in massive numbers here, together with the parasites Bird's-nest Orchid and Yellow Bird's-nest (both names referring to their extensive root system that taps into the the soil fungi and drains them of their nutrients).

Take the second trail right (signposted Le Rozier par la Jonte) and follow the track through the plantations until you arrive on the edge of the Causse on the Jonte side. The track follows the cliff's edge back towards Capluc and Le Rozier.

4 Of all the great walks one can do in the Cévennes, this stretch may very well be the most sublime. Turn after turn, the views over the Jonte Gorge and the spectacular rock pinnacles that line the cliffs, become more impressive. Griffon Vultures breed nearby and are almost constantly present, occasionally soaring by at close distance. With luck, you will spot a Golden Eagle, Black or Egyptian Vultures as well. Botanically this stretch is superb as well, although the way up on the Tarn side was more diverse.

5 The views are best somewhere in the middle of the stretch back to Le Rozier, where there is a set of massive rock 'vases' and pillars. Further ahead, you have more open views of the rocky crests above Le Rozier. Bird-watchers should scan for Blue Rock Thrush.

Following the trail automaticaly brings you back to the departure point.

Masses of Blue Aphyllanthes line the trail in the Tarn valley. This is arguably the most beautiful trails in the Cévennes.

200

Additional remarks This is a popular route, so if you want to avoid a crowd, it is best to plan this walk on week days and – in high season – to start early. The vulture observation centre, from Le Rozier two kilometres into the Gorge de la Jonte, is worth a visit (see page 211).

The 'vases' are spectacular limestone rocks encountered along the route.

Selected species of this route

Plants: Staehelina, Leuzia, Blue Aphyllanthes, Five-leaved Dorycnium, Bellflower Flax* (Linum campanulatum), Bastard Balm, Woodcock Orchid, Creeping Lady's-tresses, Branched St. Bernard's Lily, Fairy Foxglove, Alpine Mezereon, Globe-headed Rampion, Perennial Yellow-woundwort, Pyrenean Bellflower*, Lecoque's Red Valerian, Sticky Columbine, Cévennes Cinque-foil*, Pungent Pink*, Violet Limodore, Red, Dark-red, Broad-leaved Helleborines, Sessile-leaved Broom*, Liverwort, Nodding Wintergreen, Spanish Spiny Greenweed*, Pale Stonecrop, Yellow-wort, Swallow-wort, Wayfaring Tree, Wall Germander, Common Lavender
Birds: Griffon Vulture, Black Vulture, Golden Eagle, Eagle Owl, Raven, Red-billed Chough, Blue Rock Thrush, Melodious, Bonelli's and Subalpine Warblers, Firecrest
Butterflies: Blue-spot Hairstreak, Cleopatra, Sloe Hairstreak, Speckled Wood, Grayling, Fritillaries (Marbled, Silver-washed, Meadow, Heath, Queen of Spain) Scarce and Common Swallowtails, Adonis Blue, Chalkhill Blue, Long-tailed Blue, Wood White, Dingy and Large Skippers

Route 17: The schist Cévennes

4 HOURS
MODERATE

Pleasant walk through chestnut groves and
Mediterranean scrubland.
Beautiful stream with a nice flora and many
butterflies and dragonflies.

Habitats along this route: Mediterranean scrub (p. 30), heathland (p. 32), pine
forest (p. 40), chestnut grove (p. 37), river (p. 48).

The Mediterranean habitats of the eastern Cévennes are somewhat underrepresented in the Park's network of way-marked trails. This route compensates that shortcoming. In rapid succession, the most typical landscapes appear: shady streams, chestnut groves, heath land open pinewoods and Mediterranean scrubland. Particularly beautiful is the Gardon stream on the second part of the route. Visitors interested in dragonflies, butterflies and reptiles will enjoy the sun-soaked scrubland and river edges.

The Small Pincertail is one of the most frequently encountered dragonflies in the rivers of the Cévennes.

Departure point St Etienne Vallee Française

Getting there In the main street park in front of the college. From here, walk in the direction of the village centre and take the first left, direction Bureau des Postes. Go left again at the T-junction. This street leaves the village in a westerly direction. When the road forks a few hundred metres ahead take the road left, towards Auriol.

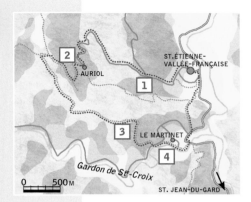

1 This minor road follows a small valley with some small cultivated plots. Mediterranean scrubland and a small stream to your left. This is a good spot for Mediterranean butterflies (in season: Southern White Admiral, Scarce Swallowtail, Lesser Purple Emperor and, more rarely, the magnificent Two-tailed Pasha). In spring you might hear (rather than see) some of the birds that are so typical for Mediterranean streams but are rare higher up in the Cévennes, like Golden Oriole and Cetti's Warbler.

2 The road crosses the streamlet and winds uphill. You are now on the cooler and moister north slope. The hillside is covered with some beautiful Chestnut groves and open meadows, which, in spring, have a nice selection of butterflies. In summer, the larger dragonfly species patrol along the road, hunting for insects. After the farmstead of Auriol, the road continues as a dirt track through the chestnut woods. The many holes in the old trunks reveal the presence of hole-nesting birds like Nuthatches, Green and Great Spotted Woodpeckers.

The Gardon is a beautiful little River, as inviting to watch dragonflies and plants as it is to take a dip on a warm summer's day.

At the next fork, follow the right track (Colombieres) and follow the yellow markings. Near the summit of the hill (where the chestnuts give way to an open forest of Maritime Pines), take a left twice and continue on the other side of the hill. Bell Heather grows between the trees here and you have good views over the Gardon. The track remains level for a few hundred metres, until a sharp turn right signals the steep descent into the valley.

3 The descent leads through a typical Mediterranean scrubland of acidic soils with all its characteristic species of bushes: Mock Privet, Strawberry Tree, Poplar-leaved and Sage-leaved Cistuses and Tree Heath. In spring, the scrubland flowers in profusion. Look out for Wolf Spider, Praying Mantis and Hooded Praying Mantis* or Empusa. Reptiles include Iberian and Common Wall Lizards, and Montpellier and Southern Smooth Snake.

The last part of the descent follows another small streamlet, before arriving on the road.

Follow the road left for some hundred metres and in the bend, take the track down to the river. Follow the trail left along the river until you arrive at the bridge.

4 The river with its smooth rocks is another highlight of this walk. Its sun-speckled water presents an irresistible invitation to spread one's towel and take a swim. The flowery margins around the rocks are perfect for butterflies (Scarce Swallowtails, Banded Grayling, Southern White Admiral) and dragonflies (Common Goldenring, Western Demoiselle, Small Pincertail and Keeled Skimmer). Between the rocks there are countless little springs, which are very much worthy of your attention. This is the habitat of a number of rare plant species including the orchid Summer's Lady's -tesses. Unfortunately we didn't find it, but they certainly ought to be around here somewhere.

At the bridge go left and take the GR track that departs to the north from the T-junction. This short track brings you back to St Etienne Vallee Francaise.

Selected species of this route

Plants: Strawberry Tree, Poplar-leaved Cistus, Tree Heath, Mock Privet
Birds: Honey Buzzard, Short-toed Eagle, Green Woodpecker, Cetti's, Melodious and Subalpine Warblers
Insects: Keeled Skimmer, Western Demoiselle, Small Pincertail, White Featherleg, Lesser Purple Emperor, Two-tailed Pasha, Southern White Admiral

---

Route 18: Orchids in the Cernon valley

4 HOURS
EASY - MODERATE

Probably the best site for orchids in the entire area.
Beautiful limestone valley with a magnificent flora.

Habitats along this route: downy oak forest (p. 35), causses (p. 49).

The legendary botanical richness of the Causses seems to reach its peak in the lovely little Cernon valley on the Causse du Larzac. In particular orchids abound here. Most species that are found scattered over the Causses, cluster together on this fortunate little hillside. On a walk in late May, it is possible to see over 20 species of orchids growing together, making it the best site for orchid watching in the Cévennes that we know. Besides orchids, there are masses of other wildflowers, a good collection of butterflies, while – as everywhere – Green Lizards lurk under the bushes and Short-toed Eagles patrol overhead to catch them.

Departure point La Cavalerie

Spectacular display of wildflowers in May, with Common Globularia, White and Hoary Rockroses and Wild Mignonette.

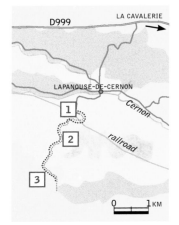

Getting there From the old N9, take the D999 west for 3.25 kms, crossing the new A75 motorway and turn left on the D562E towards Lapanouse de Cernon. Follow the road through the village until you meet the D77 and turn right (very tight turn). In just 200 metres turn left over a narrow bridge across the Cernon and follow the narrow road up to the old abandoned station and park to the left of the station only!

1 The start of the route is immediately interesting. In the field on the right side of the track opposite the abandoned railroad station, you can find Man, Green-winged, Passion-tide* (a type of Spider Orchid) and the endemic Aveyron Orchid*. From the trees, Bonelli's Warblers and Cirl Buntings sing. Look out for Tree Pipits, Firecrests and Turtle Doves as well. Check the sky here and everywhere else for Griffon Vultures and Short-toed Eagles.

The track crosses the railroad and follows it for a bit before turning left, up on the hillside just past a small cattle grid. Here it passes through limestone meadows lined by light woodland.

!
Do not park beyond the station as Aveyron Orchids grow here!

Passion-tide Orchid (left) and Yellow Bee Orchid (above) are both frequent in the Cernon valley.

PRACTICAL PART

2 The way up is a true spectacle. Drifts of White and Hoary Rock Roses flower in combination with a dazzling variety of other wildflowers (for the most eye-catching species, see box on facing page). The orchids are exceptional, both in number and in variety. In typical Cévennes fashion, there is mixture of temperate European and Mediterranean species. Butterflies are in evidence as well, particularly blues and fritillaries.

About half-way up, the track splits with one branch bearing left and the other going straight on. Follow the latter.

The endemic Aveyron Orchid is the star species of the Cernon valley.

3 Gradually, the terrain levels as you climb out of the valley. The views here are excellent and this is a good place to scan the sky for birds of prey. In the dry grasslands there are plenty of Small Spider Orchids and a few Woodcock and Yellow Bee Orchids, which were not present further down.

A little further ahead the track becomes a trail and continues in thicker woodland. This is the end of the botanical route, but you can continue further into the Downy Oak Woodland if you like.
Return over the same trail.

Additional remarks The meadows are not fenced off, but nevertheless the vegetation is sensitive to trampling. Please take utmost care when exploring the terrain!
This is a botanical trip, great for watching wildflowers and butterflies, but not suited for long distance hikers. This route is primarily of interest between late April and Mid June. After that, the flowering season of most orchids is over and the trail looses much of its special charm.

Selected species of this route

Orchids: White Helleborine, Narrow-leaved Helleborine, Violet Limodore, Lesser Butterfly Orchid, Greater Butterfly Orchid, Fragrant Orchid, Elder-flowered Orchid, Pyramidal Orchid, Early Purple Orchid, Military Orchid, Lady Orchid, Green-winged Orchid, Man Orchid, Burnt Orchid, Lizard Orchid, Small Spider Orchid, Aveyron Orchid, Early Spider Orchid, Passion-tide Orchid, Fly Orchid, Yellow Bee Orchid, Furowed Orchid, Woodcock Orchid
Hybrids: Greater x Lesser Butterfly Orchids and Lady x Military Orchids
Other plants: Pasqueflower, Pyrenean Snakeshead, Golden-drop, Hoary, White and Yellow Rockrose, Beautiful Flax, White Flax, Bastard Balm, False Vetch, Mountain Kidney-vetch, Wild Columbine, Alpine Aster, Sand-catchfly, Italian Catchfly, Grass-leaved Oxe-eye Daisy, Meadow Clary, Liverleaf, Rock Soapwort, Sessile-leaved Broom*, St. Bernard's Lily, Wild Mignonette, Pink Bindweed*, White Asphodel, Feather Grass, Horse-shoe Vetch, Star-of-Bethlehem, Weld, Common, Chalk Milkwort*, Angular Solomon's Seal
Butterflies: Marsh, Silver-washed, Marbled, Heath, Spotted, Glanville, Niobe and Queen of Spain Fritillaries. Both Swallowtails, Cleopatra, Marbled White, Black-veined White, Wood White, Speckled Wood (southern variety). Silver-studded, Chalkhill, Adonis, Large and Little Blues. Green, Blue-spot, Sloe and Ilex Hairstreaks. Duke of Burgundy Fritillary, Clouded Yellow, Berger's Clouded Yellow.

!

It can be very hot in the gorge.

Route 19: Gorge de la Vis – Cirque de Navacelles

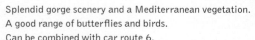

2 HOURS TO A FULL DAY
EASY

Splendid gorge scenery and a Mediterranean vegetation.
A good range of butterflies and birds.
Can be combined with car route 6.

Habitats along this route: Mediterranean scrub (p. 30) cliffs (p. 47).

This scenic walk leads deep into the Gorge de la Vis, the southernmost gorge of the region. It is rather different from the other gorges because of its denuded, Mediterranean character and because there are no villages or roads in the gorge. There is only a small trail that connects the Cirque de Navacelles to the village of St Maurice Navacelles on the Causse. Along this trail you will enjoy breathtaking views, have good chances of watching Golden Eagles, Alpine Swifts and Blue Rock Thrushes, and you can find a great variety of Mediterranean plants and butterflies, which are uncommon or absent further north.

The pretty *Philaeus chrysops* spider hunts among the rocks in the gorge.

Departure point Cirque de Navacelles (south of Blandas)

From the car park, walk through the village and follow the narrow trail past the little cemetery. It leads up the hill and over it.

1 The picturesque, abandoned village of Cirque de Navacelles is popular with tourists because of its beautiful location and the cascade of the river, where many people come to swim. For the naturalist the village is interesting for its birds (Alpine Swift, Blue Rock Thrush, Melodious Warbler, Scops Owl, if still present) and rock plants.

Take the trail past the small cemetery (Lizard Orchid grows here) that zigzags up and over the hill into the gorge. Follow this trail for as long as you like. The way back is the same as the way up. After the initial climb the trail stays level for most of the route. Only the last stretch up to St Maurice is a challenging climb again.

2 The track, partly shaded partly sunny, cuts through open scrubland, rock slopes and low, Mediterranean woodland. Along the way, look for songbirds like Cirl and Ortolan Bunting, Blue Rock Thrush, Melodious, Subalpine and Sardinian Warblers. Up in the sky look for Golden and Short-toed Eagles and Alpine Swift. The vegetation is rich in Mediterranean shrub and tree species (see boxed list). Along the trail, in spring, the yellow flowers of Dwarf Scorpion-vetch* *(Coronilla minima)* and Prostrate Toadflax are conspicuous. The large spurge here is Mediterranean Spurge, a typical species of hot, exposed slopes. The two plants of the Umbellifer family are French Sermountain* *(Laserpitium gallicum)* and Athamanta. The pretty little red spiders that creep around the rocks are the jumping spiders of the species *Philaeus chrysops*.

Due to the position and Mediterranean character of the Gorge de la Vis, there are a number of butterfly species that you will find more easily here than elsewhere in the Cevennes, like Spanish Gatekeeper, Esper's Marbled

Gorge de la Vis

White and Provence Orange-tip. There is a fair number of orchids along the track, including Woodcock, Small Spider, Yellow Bee, Lizard, Pyramidal, Lizard and Monkey Orchids. The Gorge de la Vis is reputed as one of the few sites where Large-flowered Orchid* (*Ophrys magniflora*) grows, although we never found it here. This magnificent species, formerly regarded as a large-flowered variety of the Berteloni's orchid, is definitely worth searching for.

Additional remarks Doing the trip in its entire length to St Maurice de Navacelles and back is probably too long and strenuous. However, to get a good impression of this area, it is not necessary to do the whole stretch. Even a third of the route into the gorge makes a wonderful trip.

This trip can easily be combined with the car route on the Causse Blandas (route 6).

The Etruscan Honeysuckle is one of the Mediterranean wildflowers that is common in the gorge.

Selected species of this route

Plants: Bladder Senna, Phoenician Juniper, Stinking Rue, Dwarf Scorpion-vetch, Small Spider Orchid, Beautiful Flax, Blue Lettuce, Crimean Iris, Mediterranean Spurge, Cretan Athamanta, Spanish Broom, Prostrate Toadflax

Birds: Short-toed Eagle, Golden Eagle, Griffon Vulture, Alpine Swift, Blue Rock Thrush, Sardinian, Subalpine and Melodious Warblers

Insects: Morrocan Orange-tip, Large Wall Brown, Spanish Gatekeeper, Esper's Marbled White, Cleopatra, Hooded Praying-mantis, Philaeus chrysops

Interesting sites and other extras

Donkey rides

Before Robert Louis Stevenson wrote his world famous novels 'Treasure Island' and 'The Strange case of Dr Jekyll and Mr Hyde', he published a journal of a journey called 'On a Donkey through the Cévennes'. Today, the 'Stevenson's trail' still exists. The Grand Randonnée (GR) 70 loosely follows the itinerary of the young author. It can be walked on foot, but several companies offer donkey tours with overnight stays in the mountains. Seek information at the local information centre.

Guided excursions

The LPO, the French Birdlife branch, arranges birdwatching tours in the Grands Causses. Departing from the Vulture centre just east of le Rozier, you can join a birding trip in the region. Note that these trips are not designed to locate that hard-to-find Scops Owl or Orphean Warbler. Rather, they are general field trips with birds as their topic. Information and Reservation at LPO Grands Causses, tel: 05 65 62 61 40.

Vulture watching

A trip in the Grands Causses simply isn't complete without a visit to the Belvédère des Vautours just east of Le Rozier in the Gorge de la Jonte. From the platform you can look at the colony of reintroduced Griffon Vultures through one of the telescopes of the centre. There are several cameras directed to the colony and to a feeding place from which you have live views of the vultures. The film of the reintroduction of these vultures is very interesting and amusing.

Cévennes history

The history of the Cévennes is captured in three museums:
Musée des Vállees Cévenols in St. Jean du Gard focuses primarily on the Chestnut and silk culture. www.museedescevennes.com
Musée Cévenol in Le Vigan has an impressive collection of items from everyday Cévenol life. www.musee-cevenol.com
Musée du Désert in Mialet tells the story of the Huguenots in the Cévennes. www.museedudesert.com
Centre de documentation is an impressive archive (books, references, digital documents) on the history, ethnology, geology, land use, agricultural practises, botany etc. of the Cévennes. Access is free. 3 Grand-Rue 30450 Genolhac. www.cevennes-parcnational.fr

Caves and other impressive geological features

Like most limestone regions there are some impressive geological features like caves, gorges, karsts, sudden waterfalls that are worth visiting (see geology section on page 22). Here are the most impressive ones, with reference to routes if applicable. The € sign indicates that you have to pay hard cash to see it.

Caves

1. Aven Armand. Consists of a massive, 110 metres long and 45 metres high hall. Said to be the most impressive cave in the region; close to route 2, €. www.aven-armand.com
2. Grotte de Dargilan. The pink cave, coloured by iron oxides. Dargilan is a two storey cave with a massive hall, €. www.grotte-dargilan.com
3. Grotte de la Cocalière. 1 km tour along subterranean lakes, stalactites, stalagmites and other outer worldly shapes and colours, €. www.grotte-cocaliere.com
4. Grotte de Demoiselles. Superb dripstone cave, €. www.philprod.net/la_grotte_des_demoiselles
5. Grotte de Trabuc. Largest cave of the Cévennes, €. www.grottes-de-france.com/traacc.html

Gorges

6. Gorge de la Vis. Spectacular gorge (routes 6, 19)
7. Le Rozier (route 16)
8. Cirque des Baumes (route 2)
9. Gorge de Chassezac (route 11)

Karst landscape

10. Nîmes le Vieux (route 15)
11. Montpellier le Vieux, €
12. Bois de Paiolive (route 11)

Other sites

13. Abime de Bramabiau. A sudden abyss with waterfall.
14. Les Bondons. Eroded limestone steppes (route 10).

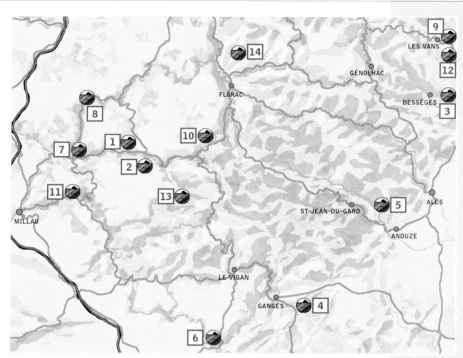

Locations of the most remarkable limestone formations (caves, karst etc.) in the region. The numbers refer to the sites mentioned on the facing page.

The karst landscape of Nimes le Vieux is freely accessible (route 15).

TOURIST INFORMATION & OBSERVATION TIPS

When to go

The Cévennes are worth visiting at any time between late April and mid October. Each period has its own attractions. In April, the flora quickly develops from the first carpets of Wood Anemones on the forest floors to the profusion of orchids and other wildflowers that characterise the Cévennes all through May and June. The summer flora is only slightly less impressive. Even up on the Mont Lozère the flowers are superb in May, with the slopes frosted white with Pheasant's-eye Daffodils and dusted yellow with broom.

Late May and June is the best time for birdwatching. In summer, birds have quietened down and only the vultures and the Short-toed Eagles are still easily found. Butterflies and reptiles appear in good numbers in the second half of May and may be encountered all through summer. From early June the Feather Grasses start to appear on the Causses, giving entire hillsides their typical, beautiful silver wash.

When August comes, the Cévennes have entered a state of lazy summer slumber. The Causses are yellow and dry, and the woodlands of the Cévennes valleys are silent and hot. At the same time, however, the crowds of holidaymakers arrive. Sluggish rows of campervans snake their way through the gorges. All in all, August is not the best time for a visit, but the Cévennes, rugged as they are, still have plenty of quiet corners; you just have to go further off the beaten track. On the generally cold and wet Mont Lozère, however, August (as well as July) is pleasant and fresh.

Impressive thunderstorms can be witnessed in summer on the Causse and the Lozère. September then brings a modest autumn flora and, of course, masses of ripe chestnuts. It is also the time of rutting Red Deer and Mouflon. In October and early November the Chestnut leaves turn to a beautiful yellow before falling and leaving the region in its winter solitude.

Accommodation

There are plenty of campsites, hotels and 'gites' (holiday homes) in the area. In some parts, many more than you'd like! Some have taken the shape of large for-the-whole-family pleasure resorts, which might not be the spots you want to end up in if you have come for the tranquillity of nature. In July and August accommodation and campsites in the Gorge du Tarn, near Florac and on the eastern edge of the Cévennes may be full or crowded.

If you can move around during your visit, it pays to pick two or three places for a multiple day stay; one in the south-central part of the Causse (to cover Mont Aigoual and the southern Causses), one on the western flanks of the Lozère (to cover the Lozère and Causse Méjean) and one in the central part of the eastern Cévennes (to cover the schist Cévennes and Bois de Païolive).

Means of transport

The car is by far the easiest way to travel in the region. The road network is extensive and well maintained, but since most of the roads are narrow and winding, driving is slow. Those with motion sickness will be put severely to the test as will those with a fear of heights. The area is also superb for the sportive cyclist. Even though the rugged terrain is challenging, to say the least, the winding routes force motorists to drive slowly, while the fine-mazed infrastructure offers a multitude of quiet roads with frequent opportunities for overnight stay. Add to this the generally shady conditions in the valleys and cycling suddenly becomes much easier.

Those using public transport have a more difficult task getting around in the Cévennes. The major towns – Millau, Mende, Villefort, Génolhac and La Grand Combe – are connected to the good French rail network, but from there you are pretty much on your own. There are very few buses connecting the villages in the Cévennes. Visit www.keolis-aveyron.com for buses around Millau, www.mende.fr for buses around Mende and www.cevennes-tourisme.fr for buses from Florac to Alès.

Permits

To enter the National Parks you do not need to pay a fee nor obtain a permit, except for attractions such as caves, museums and Montpellier le Vieux. The Mont Lozère has a rather complicated passport system for its sites. The "passport" can be obtained at the visitors' centre in Pont de Montvert, and offers entrance to museums and other attractions. However you do not need the "passport" for the walks on the Mont Lozère.

Guided tours

Between May and September, wildlife and wildflower excursions (including evening beaver watching trips) are organised at the Val de Cantobre campsite, just north of Nant in the southwestern part of the region. At the time of writing these are led by Paul Knapp, the co-author of this book. Excursions are free for visitors of the campsite, and open to outsiders for a small fee.

Maps

There is a 1:100,000 (1 cm = 1 km) tourist map available from the IGN (Institut Geographic National) that covers the entire National Park. This map, together with the maps in this guidebook, will suffice for the excursions we describe and for further general explorations. However, the IGN has also produced very good, detailed 1:25.000 maps covering smaller sections of the area. These maps are indispensable if you want to explore some parts of the Cévennes more thoroughly on foot. These maps and the aforementioned tourist map are for sale in most bookshops, tourist information centres and campsites.

Recommended reading

There are numerous general field guides available that are helpful for those wanting to find and identify the birds, plants and animals of the Cévennes.

From the generous supply of bird books we personally recommend the Collins Bird Guide by Svensson & Grant, as it has good descriptions and beautiful and accurate drawings. The Collins field guides on the reptiles and amphibians of Britain and Europe (by Nicholas, Arnold, Denys and Ovenden) and on butterflies of Britain and Europe (by Tolman and Lewington) are also of high quality. In most European countries the Collins guides are translated and distributed by local publishers under other titles. German-speaking visitors with an interest in amphibians would do well to get a copy of Die Amphibien Europas (by Nöllert & Nöllert, Frankh-Kosmos Verlag). A Dutch translation is published by Tirion. The Dutch Veldgids Amfibieën en Reptielen (by Stumpel & Strijbosch, KNNV Uitgeverij) is the one to use for Dutch visitors. Dragon-fly enthusiasts should rush to the shops and buy the excellent Field guide to the dragonflies of Britain and Europe by K.D. Dijkstra and Lewington (British Wildlife publishing).

If you master French a little, there is a wealth of information for the naturalist available. First of all, there is the Faune de Lozère (published by Alepe, ISBN 2-9514722), which covers the whole vertebrate fauna of the Lozère province (i.e. 2/3 of the Cévennes and Grands Causses region). A volume on the invertebrates is apparently on the way.

Wildflower enthusiasts should get the two very helpful guidebooks 'Fleurs et Paysages des Causses' (by Christian Bernard; ISBN 2 84156 073) and 'Flore du Parc National des Cévennes' (by the National Park; ISBN 2 84156 094 5), both published by Editions du Rouergue (www.lerouergue.com). Together they cover some 750 of the most typical wildflowers of the region with good descriptions and photographs. Being at the crossroads of the Atlantic, central European, Mediterranean and Alpine flora regions, many of the more general wildflower books can be of help, but all of them only to a limited extent. Most of the plant species covered in this guidebook can be found in our picture database (see www.crossbillguides.org). Also, the website www.floralps.com has a large number of relevant plant species.

If you are more ecologically minded, there is the very good 'Guide du Naturaliste de Causse Cévennes' with good descriptions and excellent photographs of the 55 vegetation types in the region, including lists of the typical flora and fauna encountered in them. This very good book is only locally available; we only encountered it in the

exhibition centre on Col de Serreyrède (Mont Aigoual).
The local guidebooks are for sale at the vulture centre just east of Le Rozier and at the Maison du Parc in Florac. Some of the guidebooks are for sale in the other interpretation centres and the museums in the area.

Annoyances and hazards

Other than heights and narrow winding routes (which will put any driver who is afraid of heights to the test and any passenger with motion sickness reach for their pills), there is little to worry about in the Cévennes. There are few animals that are dangerous. None of the sometimes fierce-looking insects are dangerous although, of course, wasps and hornets should be treated with caution. The large, black Carpenter Bees and the large dragonflies are, contrary to popular belief, completely harmless.

Of the snakes, only the poisonous Adder and Asp Viper are dangerous, but, unless you go looking for them, they are hardly ever encountered. Both are readily distinguishable by the zigzag pattern on their back and their mean-looking vertical pupils. In the rare case of a viper or an unidentified snake biting you, you'd best seek medical advice directly. Note that the water snakes you see are usually harmless Viperine Snakes, which are no reason for concern.

The only animal one should be wary of is the tick. Ticks are small, blood-sucking animals, related to the spiders. They occur in high grass and in oak woodlands with many deer and boar. The ticks themselves are harmless, but some carry and may transmit Lyme disease. This is easily cured, but, if untreated, it can do much long term damage. Therefore, it is important to look for suspicious tick bites. A normal tick bite causes a reddish spot around the tick, but the Lyme carrying ticks cause a white spot with a distinct reddish ring around them. Should you find this, seek medical advice.

During some summers, temperatures may rise up to 35 or 40° C, so be prepared with plenty of water and something to protect you against the sun.

Responsible Tourism

'Take nothing but your photo, leave nothing but your footprint', is the well-known phrase that summarises the idea of responsible tourism. It goes without saying that, as a visitor to a nature reserve, you have a responsibility to leave your surroundings and everything in it undisturbed. But maybe it is less obvious what is and isn't damaging in the case of the Cévennes and Grands Causses. So here is what you should be especially aware of when visiting this area.

Keep a low profile. The massive annual streams of visitors to the Cévennes and Grands Causses seem to leave the area relatively unaffected. However, there is some disturbance to nature due to noisy behaviour, picking wildflowers, making an open fire, illegal camping and such things. All of these are forbidden for obvious reasons and one should take care to abide by these rules.

The biggest threat to the nature of the Grands Causses is the abandonment of animal husbandry. The sheep that maintained the open, steppe-like plains are disappearing because they are becoming economically unviable. This is where ecotourism can help! Plenty of very good and tasty local foods (sheep's cheese in particular, but also meat and wine) are produced in the area, the purchase of which is supporting the traditional, nature enriching land use. Adherence to the good cause of nature conservation is seldom as tasty as it is in the Cévennes.

There is a good organic farm with local produce in the hamlet of Hyelzás (close to route 2), but local produce is also for sale in supermarkets and on campsites throughout the region.

Nearby destinations worth a visit

The southern half of France has many more superb sites to offer. For explorations deeper into the Mediterranean region, you should head south-east. Only a stone's throw away from Alès lies the Camargue, one of the most well-known wetlands of Europe and the obvious next stop on your travels in France. It is famed for its large Flamingo colony – the largest in Europe – and its typical grey horses and bulls. The landscape and the atmosphere of the Camargue is unique. It is also a superb site for birdwatching. Just next to the Camargue are the marshes of Marais du Vigueirat and the stony steppes of La Crau, where several bird species can be found which occur nowhere else in France. All these areas together, plus the limestone hills of the Alpilles, form one large natural area, described in the *Crossbill Guide to the Camargue, Crau and Alpilles*.

Just east of this lies the Provence, a highly scenic area with valleys and small, rocky mountain ranges with a superb flora and an enormous variety of butterflies. For the naturalist, the Provence is a collection of different sites, such as the Massif de Maures, the Grand Canyon du Verdon and Hyères, which has the highest diversity of orchids in France. Beyond the Provence, on the Italian border, lie the Maritime Alps with the National Park Mercantour, a superb mountain region on the edge of the Mediterranean (one of the 10 'biodiversity hotspots of the Mediterranean', particularly rich in wildflowers and butterflies).

You can also opt to go to the southwest, where the Pyrenees are the obvious target for many more explorations.

North of the Cévennes, you traverse the Massif Central before coming to the central French lowlands of the Loire basin. Here, the Allier River near Moulins is an interesting area to explore and – further west – the multitude of ponds in La Brenne form a famous birdwatching spot.

Finding orchids

The following list shows the habitat, soil type (see map on page 22) and flowering time of all Cévennes and Grands Causses orchid species.

ENGLISH NAME	SCIENTIFIC NAME	Habitat	Frequency	Soil
Giant Orchid	Himantoglossum robertianum	msj	r	l
Sombre Bee Orchid	Ophrys fusca / lupercalis	msj	r*	l
Dense-flowered Orchid	Neotinea maculata	msj	vr*	l, s
Large-flowered Orchid	Ophrys magniflora	msj	vr*	l
Small Spider Orchid	Ophrys araneola	c	f	l
Yellow Bee Orchid	Ophrys lutea	c, msj	r-lc	l
Woodcock Orchid	Ophrys scolopax	c, cw, pw, s, msj	f	l
Pink Butterfly Orchid	Orchis papilionacea	m, c	vr	l
Green-winged Orchid	Orchis morio	c, pw, s	c	l
Monkey Orchid	Orchis simia	c, pw, s	c	l
Early Purple Orchid	Orchis mascula	c, cw, dw, m	c	s, g, l
Provence Orchid	Orchis provincialis	dw	r*	s
Early Spider Orchid	Ophrys aranifera	c, pw	f	l
Black Spider Orchid	Ophrys incubacea	msj	r*	l
Passion-tide Orchid	Ophrys passionis	c, pw	r	l
Tongue Orchid	Serapias lingua	sp, m	vr	l, s
Lady Slipper Orchid	Cypripedium calceolus	cw	vr	l
Greater Butterfly Orchid	Platanthera chlorantha	c, cw, pw, dw	f	l
Elder-flowered Orchid	Dactylorhiza sambucina	c, m, sm	c	l, g
Early Marsh Orchid	Dactylorhiza incarnata	sp	vr	l
Western Marsh Orchid	Dactylorhiza majalis	sp	vr	l
Robust Marsh Orchid	Dactylorhiza elata	m, sp	lc	l, s
Military Orchid	Orchis militaris	c, cw, m, pw, dw	c	l
Lady Orchid	Orchis purpurea	c, cw, m, pw, dw	f	l
Man Orchid	Orchis anthropophora	c	c	l
Burnt Orchid	Orchis ustulata	c, cw, m	c	l
Lax-flowered Orchid	Orchis laxiflora	sp	vr	l
Bug Orchid	Orchis coriophora	sp, m	r	l, s
Furrowed Orchid	Ophrys sulcata	c	r	l

mar		apr		may		jun		jul		aug		sep		okt	
x	X	X	X	X	x										
x	X	X	X	x											
		x	X	X	X	X	X	X	X	x					
		x	X	X	X	x									
		x	X	X	X	x									
				x	X	X	X	X	x						
				x	X	X	X	X	x						
				x	X	X	X	x							
				x	X	X	X	X	x						
				x	X	X	X	X	x						
				x	X	X	X	X	x						
				x	X	X	X	x							
				x	X	X	X	x							
				x	X	X	X	x							
				x	X	X	X	X	x						
				x	X	X	X	x							
				x	X	X	X	X	x						
				x	X	X	X	X	x						
				x	X	X	X	x							
				x	X	X	X	x							
				x	X	X	X	x							
				x	X	X	X	X	x						
				x	X	X	X	x							
				x	X	X	X	x							
				x	X	X	X	X	X	X	x				
				x	X	X	X	x							
				x	X	X	X	X	x						
				x	X	X	X	x							

Habitat: c = causse limestone hills; cw = cliff woodland; m = meadow; sm = subalpine meadow; pw = pine woods; dw = deciduous woods; sp = springs; sc = scrub; msj = Mediterranean scrub on jurrasic limestone (=the eastern and southern rim).
Frequency: c = common; l = locally common in particular habitats; f = frequent; r = rare; vr = very rare, only in very few sites; * = presence unknown.
Soil: l = limestone; g = granite; s = schist

ENGLISH NAME	SCIENTIFIC NAME	Habitat	Frequency	Soil
Fly Orchid	Ophrys insectifera	c	f	l
Aymonin's Orchid	Ophrys aymoninii	c	lc	l
Aveyron Orchid	Ophrys aveyronensis	c	lc	l
Bee Orchid	Ophrys apifera	c, cw, dw, pw	lc	l
White Helleborine	Cephalanthera damasonium	cw, pw, dw	c	l
Narrow-leaved Helleborine	Cephalanthera longifolia	cw, pw, dw	c	l
Bird's-nest Orchid	Neottia nidus-avis	pw, dw, cw	c	l
Violet Bird's-nest	Limodorum abortivum	pw, dw, cw	f	l
Lizard Orchid	Himantoglossum hircinum	c, m, cw	c	l
Lesser Butterfly Orchid	Platanthera bifolia	c, dw, s, sp	f	l, s, g
Small-white Orchid	Pseudorchis albida	sm	r	g
Heath Spotted Orchid	Dactylorhiza maculata	m, sm, sp	lc	g, s
Common Spotted Orchid	Dactylorhiza fuchsii	cw, pw, dw	f	l
Pyramidal Orchid	Anacamptis pyramidalis	c, m	c	l
Red Helleborine	Cephalanthera rubra	cw, pw, dw	f	l
Tremol's Helleborine	Epipactis tremolsii	msj	r	l, s
Greater Twayblade	Listera ovata	pw, dw, cw	f	l
Fragrant Orchid	Gymnadenia conopsea	c, m	c	l
Frog Orchid	Coeloglossum viride	c, m, sm, s	f	l, g
Creeping Lady's-tresses	Goodyera repens	pw, cw	c	mostly l
Summer Lady's-tresses	Spiranthes aestivalis	sp	r	g
Lesser Twayblade	Listera cordata	pw	r	l, s, g
Marsh Helleborine	Epipactis palustris	sp	r	l
Coralroot Orchid	Corallorrhiza trifida	dw	vr	l, g, s
Dark-red Helleborine	Epipactis atrorubens	pw, dw, cw	f	l
Small-leaved Helleborine	Epipactis microphylla	dw	r	l, s
Broad-leaved Helleborine	Epipactis helleborine	pw, dw, cw	c	l, g, s
Violet Helleborine	Epipactis purpurata	cw	vr*	l, s
Narrow-lipped Helleborine	Epipactis leptochila	pw, dw, cw	r	l
Müller's Helleborine	Epipactis muelleri	pw, dw, cw	r	l
Ghost Orchid	Epipogium aphyllum	dw	r	g, l
Autumn Lady's-tresses	Spiranthes spiralis	c, m	r	l

	mar	apr	may		jun		jul		aug		sep		okt	
			x	X	X	x								
			x	X	X	x								
			x	X	X	x								
			x	X	X	X	x							
			x	X	X	x								
			x	X	X	x								
			x	X	X	X	x							
			x	X	X	x								
				x	X	X	x							
				x	X	X	X	x						
				x	X	X	X	x						
				x	X	X	X	x						
				x	X	X	X	x						
				x	X	X	x							
				x	X	X	x							
				x	X	X	x							
				x	X	X	X	x						
					x	X	X	X	x					
					x	X	X	X	x					
					x	X	X	X	x					
					x	X	X	X	X	x				
					x	X	X	x						
					x	X	X	x						
					x	X	X	X	x					
						x	X	X	X	x				
						x	X	X	X	x				
						x	X	X	X	x				
						x	X	X	X	x				
						x	X	X	X	x				
						x	X	X	X	x				
						x	X	X	X	x				
								x	X	X	X	x		

Habitat: c = causse limestone hills; cw = cliff woodland; m = meadow; sm = subalpine meadow; pw = pine woods; dw = deciduous woods; sp = springs; sc = scrub; msj = Mediterranean scrub on jurrasic limestone (=the eastern and southern rim).
Frequency: c = common; l = locally common in particular habitats; f = frequent; r = rare; vr = very rare, only in very few sites; * = presence unknown.
Soil: l = limestone; g = granite; s = schist

Bird list

Numbers between brackets () refer to the routes in this guidebook.

Herons Grey Heron occurs along the broader parts of the rivers, particularly around Florac. Little Egret frequent the broader Mediterranean streams just outside the Cévennes.

Ducks Very few. Only Mallard and, in winter, Teal are regular.

Vultures On the Grands Causses, Griffon Vultures are a common sight (1, 2, 4, 12, 14 15 16, 19). A large colony is present in the Gorge de la Jonte (2). Griffons can be seen cruising the entire area of the Grand Causses, but the best routes are 2 (the vulture centre!) and 14. Black Vulture is less common, but every now and then they can be found along the same routes. Egyptian vulture is, with only 1-4 pairs in the area, quite rare. They can be found on the Causse and in the rocky limestone region of the eastern Cévennes (11).

Eagles The Cévennes is a superb area for Short-toed Eagle. It occurs throughout the area, but is most abundant on the Causses and is absent from the high slopes of the Lozère and Aigoual. The Golden Eagle has become quite rare since the collapse of the rabbit population caused by Myxomatosis. They can be seen in the gorges and in the Lozère area (1, 2, 6, 16 and particularly 19).

Harriers Montagu's and Hen Harrier both occur in fairly low numbers on the Causses (2, 4, 6, 14, 15), with Montagu's being two to three times more numerous than Hen Harrier. Both species also occur on the Mont Lozère (7, 8, 9), but here the Hen Harrier is the more abundant species.

Other raptors After Short-toed Eagle, the Honey Buzzard is perhaps the second most numerous raptor of the Cévennes. It is present in wooded areas near open Causse or heathland. Buzzards are uncommon breeding birds of the Causse (mostly breeding in the wooded hillsides of the gorges) and the Mont Lozère (7, 8, 9). Both Black and Red Kites occur in very low numbers, with the latter being slightly more numerous. Black Kites however, become quite common in the Eastern low Cévennes (11) and the lowlands west of the Cévennes.
Both Goshawk and Sparrowhawk are frequent throughout the wooded areas of the Cévennes (particularly 3, 5, 7, 11, 13, 18 and 17).

Falcons Kestrel is a frequent bird of the Causses. Peregrine breeds in low numbers on steep cliffs (2, 11, 16, 19). Hobby is a rare bird in the region. Merlin occurs on the Causses in winter in small numbers.

Partridges The Red-legged Partridge is fairly common in all types of light woodland and scrub. Quail can be heard (though is, as always, difficult to see) in the cultivated areas and the steppes of the Grand Causses. It can be quite common, but the numbers vary from year to year. Pheasant has been introduced for hunting and is locally quite abundant in farmland. Grey Partridge is a very localised breeding bird on the Mont Lozère (1).

Between 1978 and 1993, Capercaillie was reintroduced in the Cévennes after 300 years of absence. It occurs in the tranquil coniferous woodlands of the north slopes of the Mont Lozère and Montagne des Bougès, but is very hard to see.

Little Bustard and Stone Curlew It is unclear whether Little Bustard still occurs in the area, but, if so, it is only present in the military terrain on Causse Larzac, which cannot be visited. Stone Curlews are quite abundant on the open parts of the Causses, particularly on Méjean. They are best located on spring evenings, when they betray their presence with their melancholy *kir-lewww* call (particularly 2, but also 4, 6 and 14).

Wading birds The Common Sandpiper is the only wader you are likely to encounter. It breeds in small numbers along the Tarn river (1). Woodcock breeds in low numbers on the Mont Lozère and the Causse de Sauveterre, but it is very difficult to find. During migration, Common Snipe and Dotterel pass over the Mont Lozère (7). Also Green Sandpiper, Black-winged Stilt and Redshank can occasionally be found at Lavognes during migration (2, 14).

Pigeons, Doves and Cuckoos Stock Dove is a fairly rare breeding bird of the river valleys. Wood Pigeon is remarkably uncommon. Collared Doves and Feral Pigeons are the most numerous of the pigeons, occurring widely in and around habitations. Turtle Doves are frequent, particularly on the Causse. Cuckoo also occurs throughout.

Owls The most easily encountered owl is Little Owl, which reaches highest densities on the Causses (particularly 2). The Scops Owl, a Mediterranean species, is rare, but can sometimes be heard calling from the larger trees in the villages in the southern part of the Cévennes (Hures and St. Pierre des Tripiers on Causse Méjean (2), Millau, Blandas (6) and Cirque de Navacelles (19). Eagle Owls are quite numerous in the gorges. An alleged good place is in the Tarn near the village of Rivière sur Tarn. They are only found with luck, or, in early spring, by following the call. There are a few pairs of Tengmalm's Owl, a northern species, present in the older forests on the Aigoual (5).

Tawny and Long-eared Owls occur in the forested areas. Barn Owl is very rare although it used to breed in Nant.

Nightjar and Swifts Nightjar is quite common on the Causses, particularly on Larzac and Noir. They can be seen over the rivers and at Lavognes at dusk (the Lavogne at Montredon (4) is a good spot).

Common Swifts breed in most, if not all, towns and villages. Alpine Swift breeds locally in the gorges (particularly 11 and 19). They are best seen from the cliff tops, rather than from below. At Point Sublime in the Tarn, they fly by at only 2 or 3 metres distance!

Kingfisher, Hoopoe, Roller and Bee-eater Kingfishers are quite rare in the Cévennes, because here they prefer fairly large rivers. Only in the Tarn (2) are they more easily spotted. In the streams in the foothills of the Cévennes they become quite numerous. The Causse is the perfect habitat for Hoopoes and it is here where you mostly find them (1, 2, 4, 6, 14 and 15), although they are not as common as the habitat might suggest. Bee-eaters are very rare in the area, but occur widely further south, west and east of the Cévennes. They occasionally breed on Causse Blandas and Larzac, and regularly in the Tarn near Aguessac just north of Millau. Bee-eaters are common in the Rhone valley. Similarly, Roller breeds occasionally on the Causse (particularly Blandas and Larzac), but is more numerous in the Mediterranean Plain.

Woodpeckers Of the woodpeckers, only Great-spotted and Green are common, occurring in any type of woodland. Black and Lesser Spotted Woodpecker breed only in mature woodland areas (5, 7), but forage regularly in younger woodlands. Wryneck prefers the more open landscape of the Causse and the Cevenol valleys, but it is thinly spread over the area.

Larks Larks occur in the open areas, primarily on the Causses and the Mont Lozère. Skylark is common in all open areas, while Woodlark is a typical Causse bird (1, 2, 4, 6, 10, 14, 15, 18). Short-toed Lark is extremely rare, if it hasn't disappeared completely. It used to breed on Causse Méjean. Any new sightings should be reported at the LPO (Ligue pour la protection des Oiseaux) , 12720 Peyreleau, Tel : 05 65626140.

Pipits and Wagtails There are four species of pipits and two wagtails in the area. Tree Pipit is common in open areas with some light woodland, particularly on the Causses (1, 2, 4, 6, 10, 14, 15, 18). Water Pipit is found only on Mont Lozère (8, 9), Meadow Pipit also on Mont Aigoual (5, 7, 8, 9). The open Causses form an excellent place for spotting the otherwise difficult to find Tawny Pipit. It is fairly common in open steppes (2, 4, 6, 14, 16).

White and Grey Wagtails are common along the streams (1, 2, 3, 5, 9, 17). White Wagtail is also found in villages and rural areas. Blue-headed (Yellow) Wagtail is a common migrant in Spring, particularly near the Lavognes (2, 4).

Swallows and Martins Of the swallows and martins, the Crag Martin is probably the most numerous, breeding in rocks, but also in old bridges and houses (1, 2, 5, 11, 14, 15, 16, 17, 19 and most villages). Barn Swallow and House Martin are found in most villages and hamlets.

Accentors and Dipper The Dunnock or Hedge Accentor is quite uncommon, breeding in park-like woods and near habitation only in the higher Cévennes. Alpine Accentors winter in fair numbers in rocky upland and in gorges, often in villages. The Mont Lozère is a good area (9) to encounter them as well as the Gorge du Tarn (2; la Malène). Dipper is a fairly common breeding bird of the streams in the area (1, 2, 5)

Redstarts, Chats, Wheatears and Nightingale Common Redstart occurs in parkland and orchards in the high and the eastern Cévennes (best on 3 and 7). Black Redstart can be found in all villages (check the roofs), karst areas and on rocky slopes. Stonechat is widespread and fairly common on the Causses and on the Lozère. In the latter area (7, 8, 9), it is accompanied by Whinchat. Migrant Whinchats can be found on the Causses. Northern Wheatear is a fairly common breeding bird in open areas, particularly on Causse Mejean and Mont Lozère. Black-eared Wheatear by contrast, is one of those Mediterranean species that is very rare in the area, but apparently does occur from time to time on the Causse Méjean, Larzac and Blandas (2, 4, 6). Nightingales are very common, occurring in scrub or woodland, particularly in damp areas.

Thrushes and Rock Thrushes Blackbird, Song Thrush and Mistle Thrush are all common in their preferred habitat. A small population of Ring Ouzel occurs on the forest edges of the Mont Lozère (7). The beautiful Rock Thrush is another sought-after bird that can be found with relative ease. It occurs in karst areas on the Causses (6, but particularly 2 and 15 are good for finding this bird). Blue Rock Thrush is much harder to find. It occurs in low numbers in the rocky gorges, particularly the warm, south-facing slopes in the Mediterranean gorges (16 and 19 are your best bets).

Warblers, Goldcrest, Firecrest Melodious Warbler is common in any shrubby terrain, but particularly on shrubby river banks (2, 5, 6, 12, 18, 19). In Mediterranean river-bank scrub (11, 19), listen for Cetti's Warbler, which is common in the Mediterranean plain and just reaches the edges of the Cévennes. (Western) Bonelli's Warbler is common, occurring in all sorts of woodland, but particularly in coniferous woods. It is one of the most common birds of the pine plantations on the Causses, and betrays its presence, particularly in spring, by its clear song of repeated tones; dzi-dzi-dzi-dzi-dzi.
No less than seven species of Sylvia warblers occur in the area. Most numerous is the familiar Blackcap, which occurs in all wooded areas with a well-developed undergrowth. Whitethroat, Garden Warbler and Subalpine Warblers occur in fair numbers in scrub-

land (1, 2, 3, 4, 6, 14, 17, 18, 19). For Dartford, Sardinian and Orphean Warblers you have to search a bit more carefully. Dartford and Sardinian Warbler are present in the hot Mediterranean scrub on the edges of the Cévennes (11, 17, 19, but in general in the entire zone of low hills on the eastern and southern rim of the Cévennes). The Orphean Warbler is usually a difficult species to find, but in the Cévennes you have a good chance of tracking it down. It is a bird of warm, open land with trees and large bushes and is locally quite common on the southern Causses. The best place is Causse Blandas (6).

Firecrest is a very numerous bird of coniferous forests. Goldcrest is less common, and is to be found in the same habitat.

Your chances of finding songbirds are increased considerably if you know the bird songs. Good CD's are available through nature book stores and the stores of Birdlife / RSPB.

Flycatchers Both Pied and Spotted Flycatcher occur, but both are quite rare. Spotted Flycatcher is found near hamlets, in orchards and on forest edges. Pied Flycatcher breeds in very low numbers in the chestnut groves and the oak woodlands of the valleys. Since most of its range lies to the north, Pied Flycatchers pass through in fair numbers during migration, and may at these times be seen at Lavognes (2, 4).

Tits Great, Coal, Crested, Blue and Long-tailed Tits are common in their preferred habitat (see your bird book). Marsh Tit is fairly common in a wide range of habitats, but seems most common on the Lozère.

Wallcreepers, Treecreepers and Nuthatch Wallcreeper winters in low numbers on karst rocks and steep rock cliffs in the gorges. Since it is such a secretive birds and since there is so much cliff to examine, it is a very difficult bird to find. A good spot for them is Balcon de Vertige in the Jonte Gorge. Scan the large pillar of rock used by a school for rock climbers. Another site is on the rocks above la Roque Sainte Marguerite in the Dourbie Gorge (To get to this site take the footpath from the bottom of the gorge and look at any suitable cliff as you climb).

Short-toed Treecreeper is a common, but secretive bird of all wooded areas, varying from orchards to forests. Eurasian Treecreeper is much rarer, occurring only in the older coniferous woods in the mountains (5, 7). Nuthatch is found throughout the older forests and chestnut groves.

Shrikes The Red-backed Shrike is a common bird of warm, open terrain with shrubs. It occurs widely on the Causses and locally on the heathlands and in rural areas in the eastern Cévennes (1, 2, 3, 4, 6, 12, 14, 15, 17 and 18). The other 3 species of shrike are quite rare. The Woodchat Shrike is a Mediterranean species that is a very rare breeding bird of the Causse, except on Causse Blandas (6), where it is surprisingly common. The

Cévennes are situated on the division line between the Great and the Southern Grey Shrikes. Formerly considered races of the same species, these birds have now been recognised as 'full' species; a distinction not all bird books have recognised. Both are quite rare, however. Southern Grey Shrike occurs on the driest, steppe-like habitats, whereas Great Grey searches out the wetter part of the Causse and also breeds on the Mont Lozère. There are no particular sites to find them, but your chances are the best along 2, 4 and 6.

Golden Oriole The Golden Oriole is a rare breeding bird of river woodlands on the edge of the region (for example at Millau).

Crows and allies Magpies, Carrion Crows and Jays are common throughout the area. Ravens are present throughout the Cévennes and can frequently be seen, particularly on the Causse and on the Lozère (1, 2, 4, 5, 6, 7, 8, 9, 14, 15, 16). Red-billed Choughs are locally common on the Causse (particularly along 2, 14 and 16). Jackdaws are common.

Sparrows Rock Sparrow is one of the birds for which birdwatchers come to the Cévennes. There are several breeding colonies on the Causse, particularly near the old hamlets on Causse Méjean (2, 14). House Sparrow is common near habitations, but Tree Sparrow is quite rare, occurring here and there on the Causse Méjean.

Finches The group of the finches is well represented in the Cévennes region. Chaffinch is common in every habitat, while Goldfinches, Greenfinches and Serins are numerous near habitations and in rural areas. Linnet occurs widely on the Causses and the Lozère. Hawfinches and Bullfinches are more difficult to find, but do occur widely in forested areas (particularly 5 and 7). A rarity in the Cévennes as a breeding bird is the Siskin, a northern and Alpine species of coniferous and alder woods. There is a small population on the Lozère (7 and near Mas de la Barque). In winter, Siskin is much more common, as is Brambling. The sought-after Citril Finch occurs near hamlets on the Mont Lozère (9 and near Mas de la Barque). Finally there is the Crossbill, which can be found in mature pinewoods (5, 7, 16).

Buntings There are five species of bunting in the Cévennes, often occurring in more or less the same habitats. Yellowhammer and Cirl Bunting are the most widespread, followed by Ortolan Bunting. The warm, shrubby Causse landscape is the perfect locality (particularly 2, 4, 6, 14, 15). Corn Bunting is fairly numerous on the open Causse (1, 2, 4, 6, 14, 15). The Rock Bunting, reputed to be common in the Cévennes, is in our experience, the rarest bunting and the hardest to find. It occurs in half open vegetation (bushes or light woodland) with rocky outcrops, and in this habitat it occurs at all altitudes both in the Cévennes and on the Causse. Listen for its high-pitched call.

GLOSSARY

Causse Old South-French name for the limestone upland plateaux in the eastern Languedoc region. The original upland is dissected by deep river gorges creating separate plateaux, each of which is called a Causse.

Col Pass in the mountains

Doline A rounded or elongated depression in limestone mountains. Dolines usually are the result from the collapse of a cave. Rainwater and sedimentation of clay particles make dolines more moist and fertile than the surrounding limestone soils. Most karst regions therefore only have crop growth in the dolines, while the surrounding land is used for extensive grazing of sheep, goats and cattle. This land-use pattern is very visible on the Causses.

Drailles Ancient drover roads for moving the cattle into and out of the mountains for the purposes of transhumance.

Endemic / Endemism A species occurring nowhere else in the world than in a given area. The Aymonin's Orchid is endemic to the Causse, meaning it occurs only on the Causse.

GR Grand Randonnee, the long distance walking trails in France.

Lavogne Small, artificial pond on the Causses, created to provide the cattle with drinking water.

Karst A collection of generally rugged landscape features (e.g. rock pinnacles, cragge soils, sudden depressions, caves etc.) associated with limestone (see text box on page 24).

Schist A type of metamorphic rock, meaning that it was transformed into rock from a former, non-rocky state (sand or sediment for example) under tectonic pressure and heat. Schist is transformed clayey sediment. Schist is typically layered rock consisting of thin scales that are pressed together, but can easily split along these layers. Schists in the Cévennes typically contain little calcium and are poor soils for agriculture.

Subalpine Vegetation zone in the mountains of temperate climates. The subalpine zone lies higher than the Montane zone and is lower than the Alpine zone. Subalpine vegetation is characterised by nearing the tree line, thus a fairly open landscape of stunted conifers and mountain heath land.

Transhumance Yearly move of cattle from the winter pastures in the lowlands (the Camargue and Crau) to the summer pastures in the mountains (the Mont Aigoual and Mont Lozère). This move over longer distances indicates a half-nomadic existence of the herders and has created a fascinating culture. Modern agriculture techniques together with the hard life associated with the biannual move has caused the transhumance tradition to dwindle, but recently, there has been a renewed interest in it, both in the Cévennes and in other parts of the Mediterranean world.

PICTURE & ILLUSTRATION CREDITS

The following numbers refer to the pages in the book. The letters refer to the position on the page: t for top, m for middle, b for bottom, l for left and r for right.

Photos

Crossbill Guides Foundation / Hilbers, Dirk: 8, 10(b) 11, 13, 14 (t+b), 17(t), 18, 20, 21, 23 (t,m,b), 27, 30, 32 (b), 35, 36 (t+b), 37, 39, 40(l+r), 41, 42, 45, 46, 47(b), 49, 50, 51(b), 52-53, 54, 55, 56, 58, 61, 62, 65, 66, 68(t), 77, 78, 79, 80, 81, 82, 83, 86, 87, 88, 89, 90(l+r), 91, 92(l+r), 94, 95, 96, 111, 117, 129, 130, 133(m+b),, 134, 136, 137, 139, 140, 144, 146, 148, 150-151, 152, 153, 157, 158, 159, 160, 163, 165 (t+b), 166, 168, 171, 172, 173, 175, 176, 177, 179, 180, 181, 182(t+b), 183, 184(b), 185, 186, 187, 189, 190, 191, 193, 194-195, 199, 200, 201, 202, 204, 205(l), 206, 208, 209, 210, 213

Crossbill Guides Foundation / Lotterman, Kim: 48, 196

Crossbill Guides Foundation / Vliegenthart, Albert: 10(r), 119, 120, 121, 122(t+b), 124(b), 125, 128, 154

Felix, Rob: 197

Juhasz, Tibor: 38

Knapp, Paul and Coulton, Suzie: 15 (m), 16, 17 (b), 40(m), 59, 70, 74, 76, 85, 98, 114, 124(t), 126, 145

Mager, Jörg: 31, 109, 131

Saxifraga / Hoogenstein, Luc: 100, 101, 133(t)

Saxifraga / Ketelaar, Robert: 43

Saxifraga / Kose, Ursa: 10(l)

Saxifraga / Knijff, Arie de: 110

Saxifraga / Kruijsbergen, Willem van: 97, 164

Saxifraga / Munsterman, Piet: 51(t), 103, 169(t)

Saxifraga / Straaten, Jan van der: 32(t), 106(t), 107, 108, 115, 116, 205(r)

Saxifraga / Winkel, Edwin: 15(t+b), 105, 169(b), 184(t)

Saxifraga / Wittgen, Ad: 102

Saxifraga / Zekhuis, Mark: 106(b), 147

Wildside Holidays / Muir, Clive: 47 (t)

World Wildlife Images / Andy & Gill Swash: 33, 44, 68, 112, 132, 214

Illustrations

All drawings, maps and illustrations by Crossbill Guides / Horst Wolter

SPECIES LIST & TRANSLATION

The following list comprises all species mentioned in this guidebook and gives their scientific, German and Dutch names. Some have an asterisk (*) behind them, indicating an unofficial name. See page 5 for more details.

Plants

English	Scientific	German	Dutch
Adenocarpus, Hairy*	Adenocarpus complicatus	Gewöhnlicher Drüsenginster	Gewone klierbrem*
Adenostyles	Adenostyles alliariae	Grauer Alpendost	Grauwe klierstijl
Adonis, Summer	Adonis aestivalis	Sommer-Adonisröschen	Zomeradonis
Alder	Alnus glutinosa	Schwarz-Erle	Zwarte els
Alyssum, Large-fruited	Hormatophylla macrocarpa	Grossfrüchtiges Steinkraut	Grootvruchtschildzaad*
Anemone, Wood	Anemone nemorosa	Busch-Windröschen	Bosanemoon
Anemone, Yellow	Anemone ranunculoides	Gelbes Windröschen	Gele anemoon
Angelica, Pyrenean	Selinum pyrenaeum	Pyrenäen-Silge	Pyreneeen selie*
Angelica, Wild	Angelica sylvestris	Wilde Engelwurz	Gewone engelwortel
Aphyllanthes, Blue	Aphyllanthes monspeliensis	Blaue Binsenlilie*	Blauwe bieslelie
Archangel, Yellow	Lamiastrum galeobdolon	Echte Goldnessel	Gele dovenetel
Arnica	Arnica montana	Arnika	Valkruid
Ash, Common	Fraxinus excelsior	Gemeine Esche	Gewone es
Ash, Narrow-leaved	Fraxinus angustifolia	Schmalblättrige Esche	Smalbladige es
Asphodel, Bog	Narthecium ossifragum	Beinbrech	Beenbreek
Asphodel, White	Asphodelus albus	Weisser Affodil	Witte affodil
Aster, Alpine	Aster alpinus cebennis	Alpen-Aster	Alpenaster
Aster, Goldilocks	Aster linosyris	Gold-Aster	Kalkaster
Athamanta, Cretan	Athamanta cretensis	Augenwurz	Athamantha*
Avens, Water	Geum rivale	Bach-Nelckenwurz	Knikkend nagelkruid
Balm, Bastard	Mellitis melissophyllum	Immenblad	Bijenblad
Basil, Wild	Clinopodium vulgare	Wirbeldost	Borstelkrans
Bastard-Toadflax	Thesium divaricatum	Sparriges Leinblatt	Wijdvertakt bergvlas
Beak-sedges	Rhynchospora sp.	Schnabelriede	Snavelbies
Bearberry	Arctostaphylus uva-ursi	Echte Bärentraube	Berendruif
Bedsraw, Round-leaved*	Galium rotundifolium	Rundblättriges Labkraut	Rondbladig walstro*
Bedstraw, Heath	Galium saxatile	Harzer Labkraut	Liggend walstro
Beech	Fagus sylvatica	Buche	Beuk

Bellflower, French	Campanula recta	Französische Glockenblume	Frans klokje*
Bellflower, Nettle-leaved	Campanula trachelium	Nesselblättrige Glockeblume	Ruig klokje
Bellflower, Peach-leaved	Campanula persicifolia	Pfirschblättrige Glockenblume	Prachtklokje
Bellflower, Pyrenean	Campanula speciosa	Pyrenäen-Glockenblume	Pyreneeen klokje
Bilberry	Vaccinium myrtillus	Blaubeere	Blauwe bosbes
Bilberry, Bog	Vaccinium uliginosum	Rauschbeere	Rijsbes
Bindweed, Pink	Convolvulus cantabrica	Kantabrische Winde	Kantabrische winde
Bird's-nest, Yellow	Monotropa hypopitys	Fichtenspargel	Stofzaad
Bird-in-a-bush	Corydalis solida	Gefingerte Lerchensporn	Vingerhelmbloem
Bistort	Persicaria bistorta	Schlangen-Knötterich	Adderwortel
Bitter-cress, Five-leaved	Cardamine pentaphyllos	Finger-Zahnwurz	Vijfbladig tandkruid
Bitter-cress, Large	Cardamine amara	Bitteres Schaumkraut	Bittere veldkers
Bitter-cress, Seven-leaved	Cardamine heptaphylla	Fieder Zahnwurz	Geveerd tandkruid
Bladderwort	Utricularia sp.	Wasserschlauch	Blaasjeskruid
Box	Buxus sempervirens	Buchsbaum	Buxus
Bramble	Rubus sp.	Brombeere	Braam
Broom, Butcher's	Ruscus aculeatus	Stechender Mäusedorn	Stekelige muizendoorn
Broom, Common	Cytisus scoparius	Besenginster	Gewone brem
Broom, Horrid Hedgehog*	Echinospartum horridum	Schreklige Igelginster*	Stekende egelbrem*
Broom, Piorno	Cytisus oromediterraneus	Piorno-Ginster*	Piornobrem*
Broom, Scorpion*	Genista scorpius	Skorpions-Ginster	Schorpioenbrem*
Broom, Sessile-leaved*	Cytisophyllum sessilifolium	Blattstielloser Geissklee	Bladsteelloze brem*
Broom, Silver-lined*	Argyrolobium zanonii	Silberginster	Zilverbrem*
Broom, Spanish	Spartium junceum	Pfriemenginster	Bezemstruik
Broomrape, Greater	Orobanche rapum-genistae	Ginster-Sommerwurz	Grote bremraap
Bryony, Black	Tamus communis	Gemeine Schmerwurz	Spekwortel
Buckthorn, Mediterranean	Rhamnus alaternus	Immergrüner Kreuzdorn	Altijdgroene wegedoorn*
Bugle, Blue	Ajuga genevensis	Genfer Günsel	Harig zenegroen
Bugle, Common	Ajuga reptans	Kriechender Günsel	Kruipend zenegroen
Burnet, Great	Sanguisorba officinalis	Grosser Wiesenknopf	Grote pimpernel
Butterbur, White	Petasites albus	Weisse Pestwurz	Wit Hoefblad
Buttercup, Aconite-leaved	Ranunculus aconitifolius	Eisenhutblättriger Hahnenfuss	Monnikskapranonkel
Buttercup, Grass-leaved	Ranunculus gramineus	Grasblättriger Hahnenfuss	Grasbladige boterbloem*
Calamint, Large-flowered	Calamintha grandiflora	Grossblütige Bergminze	Grote steentijm
Calamint, Wood	Calamintha menthifolium	Wald-Bergminze	Bergsteentijm
Candytuft, Annual	Iberis pinnata	Gefiederte Schleifenblume	Geveerde scheefbloem*
Candytuft, Prost's*	Iberis prostii	Prosts Schleifenblume*	Prosts scheefbloem*

PICTURE & ILLUSTRATION CREDITS

Candytuft, Rock*	Iberis saxatilis	Felsen-Schleifenblume	Rots-scheefbloem*
Caraway, Whorled	Carum verticillatum	Quirl-Kümmel	Kranskarwij
Cardabelle	Carlina acanthifolia	Golddistel	Gouddistel*
Carduncellus	Carduncellus mitissimus	Carduncellus*	Carduncellus*
Catananche, Blue	Catananche caerulea	Blaue Rasselblume	Blauwe strobloem
Catchfly, Italian	Silene italica	Italienisches Leimkraut	Italiaanse silene
Catchfly, Sand	Silene conica	Kegel-Leimkraut	Kegelsilene
Catchfly, Sticky	Lychnis viscaria	Gewöhnliche Pechnelke	Rode pekanjer
Chamomile, Rock*	Anthemis cretica	Kretische Hundskamille	Kretenzer kamille*
Cherry, Cornelian	Cornus mas	Kornelkirsche	Gele kornoelje
Chestnut, Sweet	Castanea sativa	Edelkastanie	Tamme kastanje
Cinquefoil, Cévennes*	Potentilla caulescens cebennis	Cevennes Kalk-Fingerkraut	Cevennes ganzerik*
Cinquefoil, Golden	Potentilla aurea	Gold-Fingerkraut	Gulden ganzerik
Cinquefoil, Marsh	Potentilla palustris	Sumpf-Blutauge	Wateraardbei
Cistus, Laurel-leaved	Cistus laurifolius	Lorbeerblättrige Zistrose.	Laurierbladig zonneroosje*
Cistus, Poplar-leaved	Cistus populifolius	Pappelblättrige Zistrose	Populierbladig zonneroosje*
Cistus, Sage-leaved	Cistus salvifolius	Salbeiblättriges Zistrose	Saliebladig zonneroosje
Clary, Meadow	Salvia pratensis	Wiesen-Salbei	Veldsalie
Clover, Alpine	Trifolium alpinum	Alpen-Klee	Alpenklaver
Clover, Crimson	Trifolium incarnatum	Inkarnat-Klee	Inkarnaatklaver
Columbine, Common	Aquilegia vulgaris	Gewöhnliche Akelei	Wilde akelei
Columbine, Sticky*	Aquilegia viscosa	Klebrige Akelei	Kleverige akelei*
Coralroot	Corallorhiza trifida	Korallenwurz	Koraalwortel
Corn-cockle	Agrostemma githago	Kornrade	Bolderik
Cornflower	Centaurea cyanus	Kornblume	Korenbloem
Cornsalad, Broad-fruited	Valerianella dentata	Gezähnter Feldsalat	Getande veldsla
Cornsalad, Crowned*	Valerianella coronata	Bekrönter Feldsalat	Kroontjesveldsla
Cornsalad, Narrow-fruited	Valerianella rimosa	Gefurchter Feldsalat	Geoorde veldsla
Cotoneaster, Wooly	Cotoneaster nebrodensis	Filzige Zwergmispel	Viltige dwergmispel
Cottongrass, Common	Eriophorum angustifolium	Schmalblättriges Wollgras	Veenpluis
Cottongrass, Hare's-tail	Eriophorum vaginatum	Scheiden-Wollgras	Eenarig wollegras
Cowberry	Vaccinium vitis-idaea	Preiselbeere	Rode bosbes
Cowslip	Primula veris	Echte Schlüsselblume	Gulden sleutelbloem
Cow-wheat, Crested	Melampyrum cristatum	Kamm-Wachtelweizen	Kamzwartkoren
Cow-wheat, Wood	Melampyrum nemorosum	Hain-Wachtelweizen	Schaduwhengel
Cranberry, Small	Vaccinium microcarpum	Kleinfrüchtige Moosbeere	Kleine veenbes
Crocus, Purple	Crocus vernus	Frühlings-Krokus	Bonte krokus

Daffodil, Pheasant's-eye	Narcissus poeticus	Weisse Narzisse	Witte narcis
Daffodil, Wild	Narcissus pseudonarcissus	Gelbe Narzisse	Wilde narcis
Daisy, Causses Oxe-eye*	Leucanthemum subglaucum	Causses Margrite*	Causses margriet*
Daisy, Grass-leaved Oxe-eye*	Leucanthemum graminifolium	Grasblättrige Margerite	Grasbladige margriet*
Dorycnium, Five-leaved	Dorycnium pentaphyllum	Gewöhnlicher Backenklee	Vijfbladige struikklaver*
Dutchman's Pipe	*See Bird's-nest, Yellow*		
Eryngo, Field	Eryngium campestre	Feld-Mannstreu	Echte kruisdistel
Everlasting, Mountain	Antennaria dioica	Gewöhnliches Katzenpfötchen	Rozenkransje
False-helleborine, White	Veratrum album	Weisser Germer	Witte nieswortel
Fern, Bracken	Pteridium aquilinum	Adlerfarn	Adelaarsvaren
Fern, Hard	Blechnum spicant	Rippenfarn	Dubbelloof
Fern, Parsley	Cryptogramma crispa	Rollfarn	Gekroesde rolvaren
Fern, Rustyback	Ceterach officinarum	Milzfarn	Schubvaren
Fir	Abies sp.	Tannen	Zilverspar
Flax, Beautiful	Linum narbonense	Französischer Lein	Frans vlas
Flax, Bellflower*	Linum campanulatum	Glocken-Lein	Klokjesvlas*
Flax, White	Linum suffruticosum	Strauchiger Lein	Wit vlas*
Flower, Garland	Daphne cneorum	Rosmarin-Seidelbast	Rozemarijnpeperboompje
Forget-me-not, Water	Myosotis scorpioides	Sumpf-Vergissmeinnicht	Moeras vergeet-mij-nietje
Foxglove	Digitalis purpurea	Roter Fingerhut	Gewoon vingerhoedskruid
Foxglove, Fairy	Erinus alpinus	Alpenbalsam	Alpenbalsem
Foxglove, Yellow	Digitalis lutea	Kleinblütiger Fingerhut	Geel vingerhoedskruid
Fumana, Heath	Fumana ericoides	Grosses Nadelröschen	Groot dwergzonneroosje*
Gentian, Cross-leaf	Gentiana cruciata	Kreuz-Enzian	Kruisbladgentiaan
Gentian, Fringed	Gentianella ciliata	Fransen-Enzian	Franjegentiaan
Gentian, Marsh	Gentiana pneumonanthe	Lungen-Enzian	Klokjesgentiaan
Gentian, Trumpet	Gentiana clusii	Kalk-Enzian	Grootbloemige gentiaan
Gentian, Yellow	Gentiana lutea	Gelber Enzian	Gele gentiaan
Germander, Golden Felty	Teucrium aureum	Gold-Gamander	Gouden viltgamander*
Germander, Mountain	Teucrium montanum	Berg-Gamander	Berggamander
Germander, Rouyan's Felty*	Teucrium rouyanum	Rouyan's Gamander*	Rouyans viltgamander*
Germander, Wall	Teucrium chamaedrys	Echter Gamander	Echte gamander
Giant-scabious, Yellow-flowered*	Cephalaria leucantha	Weissblütiger Schuppenkopf	Wit vals duifkruid*
Gladiolus, Field	Gladiolus italicus	Saat Siegwurz	Italiaanse gladiool
Globeflower	Trollius europeus	Trollblume	Europese trollius
Globularia, Common	Globuaria punctata	Gewöhnliche Kugelblume	Gewone kogelbloem
Golden-drop	Onosma fastigiata	Ligurische Lotwurz	Ligurische gouddruppel*

Golden-saxifrage, Opposite-leaved	Chrysosplenium oppositi-folium	Gegenblättriges Milzkraut	Paarbladig goudveil
Gooseberry	Ribes uva-crispa	Stachelbeere	Kruisbes
Grass, Feather	Stipa pennata	Echtes Federgras	Gewoon vedergras*
Grass-of-Parnassus	Parnassia palustris	Herzblatt	Parnassia
Greenweed, Hairy	Genista pilosa	Behaarter Ginster	Kruipbrem
Greenweed, Spanish Spiny*	Genista hispanica	Spanischer Ginster	Spaanse brem
Greenweed, Spiny	Genista anglica	Englischer Ginster	Stekelbrem
Greenweed, Winged	Chamaespartium saggitale	Flügelginster	Pijlbrem
Gromwell, Purple	Lithospermum purpuro-coerulea	Blauer Steinsame	Blauw parelzaad
Ground-pine	Ajuga chamaepytis	Gelber Günsel	Akkerzenegroen
Hare's-ear, Buttercup*	Bupleurum ranunculoides	Hahnenfuss-Hasenohr	Ranonkelgoudscherm*
Hawk's-beard, Fleabane-leaved*	Crepis conyzifolia	Grosskopfiger Pippau	Alantbladig streepzaad*
Hawk's-beard, Marsh	Crepis paludosa	Sumpf-Pippau	Moerasstreepzaad
Heath, Besom	Erica scoparia	Besen-Heide	Bezemdophei
Heath, Tree	Erica arborea	Baumheide	Boomhei
Heather	Calluna vulgaris	Heidekraut	Struikhei
Heather, Bell	Erica cinerea	Grau-Heide	Rode dophei
Helleborine, Broad-leaved	Epipactis helleborine	Breitblättrige Stendelwürz	Brede wespenorchis
Helleborine, Dark-red	Epipactis atrorubens	Braunrote Stendelwürz	Bruinrode wespenorchis
Helleborine, Marsh	Epipactis palustris	Echte Sumpfwurz	Moeraswespenorchis
Helleborine, Müller's	Epipactis muelleri	Müllers Stendelwurz	Geelgroene wespenorchis
Helleborine, Narrow-leaved	Cephalanthera longifolia	Schwertblättriges Wald-vöglein	Wit bosvogeltje
Helleborine, Narrow-lipped	Epipactis leptochila	Schmallippige Stendelwurz	Smallippige wespenorchis
Helleborine, Red	Cephalanthera rubra	Rotes Waldvöglein	Rood bosvogeltje
Helleborine, Small-leaved	Epipactis microphyla	Kleinblättrige Stendelwurz	Kleinbladige wespenorchis
Helleborine, Tremol's*	Epipactis tremolsii	Tremols-Stendelwurz*	Tremols' wespenorchis*
Helleborine, Violet	Epipactis purpurata	Violette Stendelwurz	Paarse wespenorchis
Helleborine, White	Cephalanthera damasonium	Weisses Waldvöglein	Bleek bosvogeltje
Herb-Paris	Paris quadrifolia	Einbeere	Eenbes
Hollowroot	Corydalis cava	Hohler Lerchensporn	Holwortel
Honeysuckle, Etruscan	Lonicera etrusca	Toskanischen Geissblatt	Toscaanse kamperfoelie*
Honeysuckle, Evergreen	Lonicera implexa	Windende Geissblatt	Altijdgroene kamperfoelie*
Houseleek, Cobweb	Sempervivum ararchno-ideum	Spinnweb-Hauswurz	Spinnenweb huislook
Houseleek, Common	Sempervivum tectorum	Dach-Hauswurz	Donderblad

Houseleek, Mountain	Sempervivum montanum	Berg-Hauswurz	Berg huislook
Iris, Crimean	Iris lutescens	Gelbliche Schwertlilie	Gele dwerglis*
Ironwort, Hyssop-leaved*	Sideritis hyssopifolia	Ysopblättriges Gliedkraut	Hyssopbladig ijzerkruid*
Ironwort, Simplebeak	Sideritis romana	Römisches Gliedkraut	Romeins ijzerkruid
Jasmine, Wild	Jasminum fruticans	Wilder Jasmin	Wilde jasmijn
Jerusalem-sage, Purple	Phlomis purpurea	Purpur-Brandkraut	Paars brandkruid*
Jonquil, Common	Narcissus jonquilla	Jonquille	Jonquille
Jonquil, Rush-leaved	Narcissus assoanus	Binsenblättriges Narzisse*	Rusbladige narcis*
Juniper, Phoenician	Juniperus phoenicea	Phönizischer Wacholder	Phoenicische jeneverbes
Kernera	Kernera saxatilis	Kugelschötchen	Bolhauwtje
Kidney-vetch	Anthyllis vulneraria	Gewöhnlicher Wundklee	Wondklaver
Kidney-vetch, Mountain	Anthyllis montana	Berg-Wundklee	Bergwondklaver
Knapweed, Crested*	Centaurea pectinata	Kamm-Flockenblume	Kamcentaurie*
Knapweed, Spotted	Centaurea maculosa	Gefleckte Flockenblume	Gevlekte centaurie*
Lady's-mantle, Alpine	Alchemilla alpina	Alpen-Frauenmantel	Alpenvrouwenmantel
Lady's-tresses, Autumn	Spiranthes spiralis	Herbst-Drehwurz	Herfstschroeforchis
Lady's-tresses, Creeping	Goodyera repens	Netzblatt	Dennenorchis
Lady's-tresses, Summer	Spiranthes aestivalis	Sommer-Drehwurz	Zomerschroeforchis
Larch	Larix sp.	Lärchen	Larix
Larkspur, Common	Consolida ajacis	Garten-Rittersporn	Valse ridderspoor
Larkspur, Eastern	Consolida orientalis	Orientalischer Rittersporn	Orientaalse ridderspoor*
Larkspur, Forking	Consolida regalis	Acker-Rittersporn	Wilde ridderspoor
Laurustinus	Viburnum tinus	Immergrüner Schneeball	Altijdgroene sneeuwbal
Lavander, French	Lavendula stoechas	Schopf-Lavendel	Kuiflavendel
Lavender, Common	Lavandula angustifolia	Schmalblättriger Lavendel	Echte lavendel
Leek, Alpine	Allium victorialis	Allermannsharnisch	Alpenlook
Leek, Round-headed	Allium sphaerocephalon	Kugel-Lauch	Kogellook
Leek, Yellow	Allium flavum	Gelber Lauch	Gele look
Leopardsbane, Austrian	Doronicum austriacum	Österreichische Gemswurz	Oostenrijkse voorjaars-zonnebloem*
Lettuce, Mountain	Lactuca perennis	Blauer Lattich	Blauwe sla
Lettuce, Purple	Prenanthes purpurea	Hasenlattich	Hazensla
Leuzia	Leuzia conifera	Kegel-Flockenblume*	Kegelcentaurie*
Lily, Branched St. Bernhard's	Anthericum ramosum	Ästige Graslilie	Vertakte graslelie
Lily, Martagon	Lilium martagon	Türkenbund-Lilie	Turkse lelie
Lily, May	Maianthemum bifolium	Schattenblume	Dalkruid
Lily, St. Bernards	Anthericum liliago	Astlose Graslilie	Grote graslelie
Lily, St. Bruno's	Paradisea liliastrum	Trichterlilie	Paradijslelie
Limodore, Violet	see Orchid, Violet Bird's-nest		

Linden	Tilia sp.	Linden	Linde
Liverleaf	Hepatica nobilis	Leberblümchen	Leverbloempje
Lousewort, Crested	Pedicularis comosa	Schopfiges Läusekraut	Kuifkartelblad*
Lousewort, Marsh	Pedicularis palustris	Sumpf-Läusekraut	Moeraskartelblad
Lungwort, Cévennes	Pulmonaria cebennis	Cevennes Lungenkraut*	Cevennes longkruid*
Madder, Wild	Rubia peregrina	Wilder Krapp	Wilde meekrap
Maple	Acer sp.	Ahorne	Esdoorn
Marigold, Marsh	Caltha palustris	Sumpf-Dotterblume	Dotterbloem
Marjoram, Wild	Origanum vulgare	Dost	Wilde marjolein
Melick, Hairy	Melica ciliata	Wimper-Perlgras	Wimperparelgras
Mercury, Dog's	Mercurialis perennis	Wald-Bingelkraut	Bosbingelkruid
Mespilus, Snowy	Amelanchier ovalis	Felsenbirne	Europese krentenboompje
Mezereon, Alpine	Daphne alpina	Alpen-Seidelbast	Alpenpeperboompje
Mignonette, Jacquini's*	Reseda jacquinii	Jacquini's Resede*	Jacquini's reseda*
Mignonette, Sesame*	Sesamoides pygmaea	Sternresede	Franse schijnreseda
Mignonette, Wild	Reseda lutea	Gelbe Resede	Wilde reseda
Milkwort, Chalk*	Polygala calcarea	Kalk-Kreuzblume	Kalkvleugeltjesbloem
Milkwort, Heath	Polygala serpyllifolia	Quendelblättrige Kreuz- blume	Liggende vleugeltjesbloem
Mint, Apple	Mentha suaveolens	Rundblätrige Minze	Wollige munt
Monkshood, Napels'*	Aconitum napellus	Blauer Eisenhut	Blauwe monnikskap
Moor-grass, Purple	Molinia caerulea	Blaues Pfeifengras	Pijpenstro
Moschatel	Adoxa moschatellina	Moschuskraut	Muskuskruid
Mulberry, Black	Morus nigra	Schwarze Maulbeere	Zwarte moerbei
Mulberry, White	Morus alba	Weisse Maulbeere	Witte moerbei
Mustard, Buckler	Biscutella laevigata	Glatt-Brillenschötchen	Glad brilkruid
Navelwort	Umbilicus rupestris	Felsen-Nabelkraut	Rotsnavelkruid
Oak, Downy	Quercus pubescens	Flaum-Eiche	Donzige eik
Oak, Holm	Quercus ilex	Stein-Eiche	Steeneik
Oak, Kermes	Quercus coccifera	Kermes-Eiche	Hulsteik
Oak, Sessile	Quercus petraea	Trauben-Eiche	Wintereik
Odontites, Cevennes*	Odontitis cebennensis	Cevennes-Zahntrost*	Cevennes ogentroost
Odontites, Yellow	Odontites lutea	Gelber Zahntrost	Gele ogentroost
Orchid, Aveyron*	Ophrys aveyronensis	Aveyron-Ragwurz	Aveyronorchis*
Orchid, Aymonin's*	Ophrys aymoninii	Aymonins Ragwurz	Causses vliegenorchis
Orchid, Bee	Ophrys apifera	Bienen-Ragwurz	Bijenorchis
Orchid, Berteloni's	Ophrys bertolonii	Bertolonis Ragwurz	Zadelorchis
Orchid, Bird's-nest	Neottia nidus-avis	Nestwurz	Vogelnestje
Orchid, Black Spider	Ophrys incubacea / Ophrys atrata	Schwarze Ragwurz	Zwarte spinnenophrys

Orchid, Bug	Orchis coriophora	Wanzen-Knabenkraut	Wantsenorchis
Orchid, Burnt	Orchis ustulata	Brand-Knabenkraut	Aangebrande orchis
Orchid, Common Spotted	Dactylorhiza fuchsii	Fuchs' Knabenkraut	Bosorchis
Orchid, Coralroot	*see Coralroot*		
Orchid, Dense-flowered	Neotinea maculata	Keusorchis	Nonnetjesorchis
Orchid, Early Marsh	Dactylorhiza incarnata	Fleischfarbenes Knaben-kraut	Vleeskleurige orchis
Orchid, Early Purple	Orchis mascula	Mannliches Knabenkraut	Mannetjesorchis
Orchid, Early Spider	Ophrys aranifera	Spinnen-Ragwurz	Spinnenorchis
Orchid, Elder-flowered	Dactylorhiza sambucina	Holunder-Knabenkraut	Vlierorchis
Orchid, Fly	Ophrys insectifera	Fliegen-Ragwurz	Vliegenorchis
Orchid, Fragrant	Gymnadenia conopsea	Mücken-Händelwurz	Grote muggenorchis
Orchid, Frog	Coeloglossum viride	Grüne Hohlzunge	Groene nachtorchis
Orchid, Furrowed	Ophrys sulcata	Gefurchter Ragwurz*	Groeforchis*
Orchid, Ghost	Epipogium aphyllum	Widerbart	Spookorchis
Orchid, Giant	Himantoglossum robertianum	Roberts Mastorchis	Hyacinthorchis
Orchid, Greater Butterfly	Platanthera chlorantha	Berg-Waldhyazinthe	Bregnachtorchis
Orchid, Green-winged	Orchis morio	Kleines Knabenkraut	Harlekijn
Orchid, Heath Spotted	Dactylorhiza maculata	Geflecktes Knabenkraut	Gevlekte orchis
Orchid, Lady	Orchis purpurea	Purpur-Knabenkraut	Purperorchis
Orchid, Lady's Slipper	Cypripedium calceolus	Frauenschuh	Vrouwenschoentje
Orchid, Large-flowered	Ophrys magniflora	Grossblütiger Ragwurz	Grootbloemige spiegel-orchis*
Orchid, Lax-flowered	Orchis laxiflora	Lockerblütiges Knaben-kraut	IJle moerasorchis
Orchid, Lesser Butterfly	Platanthera bifolia	Weisses Breitkölbchen	Welriekende nachtorchis
Orchid, Lizard	Himantoglossum hircinum	Bocksorchis	Bokkenorchis
Orchid, Man	Aceras antropophorum	Fratzenorchis	Poppenorchis
Orchid, Military	Orchis militaris	Helm-Knabenkraut	Soldaatje
Orchid, Monkey	Orchis simia	Affen-Knabenkraut	Aapjesorchis
Orchid, Passion-tide	Ophrys passionis	Oster-Ragwurz	Paasorchis*
Orchid, Pink Butterfly	Orchis papilionacea	Schmetterlings-Knabenkraut	Vlinderorchis
Orchid, Provence	Orchis provincialis	Provence-Knabenkraut	Stippelorchis
Orchid, Pyramidal	Anacamptis pyramidalis	Hundswurz	Hondskruid
Orchid, Robust Marsh	Dactylorhiza elata	Hohes Knabenkraut	Grote rietorchis
Orchid, Small Spider	Ophrys araneola	Kleine Spinnen-Ragwurz	Vroege spinnenorchis
Orchid, Small-white	Pseudorchis albida	Weisszüngel	Witte muggenorchis
Orchid, Sombre Bee	Ophrys fusca	Braune Ragwurz	Bruine orchis

SPECIES LIST & TRANSLATION

Orchid, Tongue	Serapias lingua	Einschwieliger Zugen-stendel	Gewone tongorchis
Orchid, Violet Bird's-nest	Limodorum abortivum	Violetter Dingel	Paarse aspergeorchis
Orchid, Western Marsh	Dactylorhiza majalis	Breitblättriges Knabenkraut	Brede orchis
Orchid, Woodcock	Ophrys scolopax	Schnepfen-Ragwurz	Snippenorchis
Orchid, Yellow Bee	Ophrys lutea	Gelbe Ragwurz	Gele orchis
Orlaya	Orlaya grandiflora	Grossblütiger Breitsame	Straalscherm
Orpine	Sedum telephium	Grosse Fetthenne	Hemelsleutel
Pansy, Wild	Viola tricolor	Wildes Stiefmütterchen	Driekleurig viooltje
Pasqueflower	Pulsatilla vulgaris	Gewöhnliche Kuhschelle	Gewoon wildemanskruid
Pasqueflower, Dark-red*	Pulsatilla rubra	Rote Kuhschelle	Roodbloemig wildemanskruid*
Pasqueflower, Spring	Pulsatilla vernalis	Frühlings-Kuhschelle	Vroeg wildemanskruid*
Pea, Everlasting	Lathyrus sylvestris	Wald-Platterbse	Boslathyrus
Pea, Spring	Lathyrus vernus	Frühlings-Platterbse	Voorjaarslathyrus
Peony, Wild	Paeonia officinalis	Echte Pfingstrose	Wilde pioenroos
Pheasant's-eye	Adonis annua	Herbst-Adonisröschen	Herfstadonis
Pheasant's-eye, Large	Adonis flammea	Flammen-Adonisröschen	Kooltje vuur
Pheasant's-eye, Yellow	Adonis vernalis	Frühlings-Adonisröschen	Voorjaarsadonis
Pine, Austrian	Pinus nigra	Schwarzkiefer	Zwarte den
Pine, Maritime	Pinus pinaster	Strandkiefer	Zeeden
Pine, Scots	Pinus sylvestris	Waldkiefer	Grove den
Pink, Granite*	Dianthus graniticus	Granit-Nelke*	Granietanjer*
Pink, Maiden	Dianthus deltoides	Heide-Nelke	Steenanjer
Pink, Montpellier	Dianthus monspessulanus	Französische Nelke	Montpellier anjer
Pink, Proliferous	Petrorhagia prolifera	Sprossende Felsennelke	Slanke mantelanjer
Pink, Pungent*	Dianthus pungens	Stechende Nelke	Stekende anjer*
Pink, Sequier's	Dianthus seguieri	Busch-Nelke	Sequiers anjer*
Pink, Wood	Dianthus sylvestris	Stein-Nelke	Bosanjer*
Plant, Curry*	Helichrysum stoechas	Mittelmeer-Strohblume	Mediterrane strobloem
Plantain, Grass-leaved*	Plantago holosteum	Gekielter Wegerich	Grasweegbree*
Polypody, Common	Polypodium vulgare	Gewöhnlicher Tüpfelfarn	Gewone eikvaren
Poppy, Common	Papaver rhoeas	Klatsch-Mohn	Grote klaproos
Privet, Mock	Phillyrea latifolia	Breitblättrige Steinlinde	Breedbladige steenlinde
Rampion, French	Phyteuma gallicum	Französische Teufelskralle*	Franse rapunzel*
Rampion, Globe-headed	Phyteuma hemisphaericum	Halbkugelige Teufelskralle	Halfbolrapunzel
Rampion, White	Phyteuma spicatum	Ährige Teufelskralle	Witte rapunzel
Rock-cress, Cévennes*	Arabis cebennensis	Cevennes-Gänsekresse*	Cevennes scheefkelk*
Rock-Jasmine, Greater*	Androsace maxima	Acker-Mannsschild	Groot mansschild*
Rockrose, Hoary	Helianthemum canum	Graues Sonnenröschen	Klein zonneroosje*

Rockrose, White	Helianthemum appeninum	Weisses Sonnenröschen	Wit zonneroosje
Rockrose, Yellow	Helianthemum nummularium	Gelbes Sonnenröschen	Geel zonneroosje
Rose, Burnet	Rosa pimpinelifolia	Bibernell-Rose	Duinroosje
Rosemary	Rosmarinus officinalis	Rosmarin	Rozemarijn
Rue, Narrow-leaved*	Ruta angustifolia	Schmalblättrige Raute	Smalbladige wijnruit*
Rue, Stinking	Ruta graveolens	Weinraute	Wijnruit
Rue, Wall	Asplenium ruta-muraria	Mauerraute	Muurvaren
Rush, Heath	Juncus squarrosus	Sparrige Binse	Trekrus
Sage, Ethiopian	Salvia aethiopis	Mohren-Salbei	Moorse salie*
Sage, Wood	Teucrium scorodonia	Salbei-Gamander	Valse salie
Sainfoin, False	Vicia onobrychoides	Esparsetten-Wicke	Esparcette wikke*
Sandwort, Cluster-flowered*	Arenaria aggregata	Büschel-Sandkraut	Dichtbloemige zandmuur*
Sandwort, Flax-flowered*	Minuartia capillacea	Feinblättrige Miere	Fijnbladige veldmuur*
Sandwort, Hairy*	Arenaria hispida	Hariges-Sandkraut*	Behaarde zandmuur*
Sandwort, Large-flowered*	Arenaria montana	Berg-Sandkraut*	Grote zandmuur*
Saxifrage, Cévennes*	Saxifraga cebennis	Cevennes-Steinbrech*	Cevennes steenbreek*
Saxifrage, Clusius'	Saxifraga clusii	Clusius-Steinbrech	Clusius' steenbreek*
Saxifrage, Livelong	Saxifraga paniculata	Trauben-Steinbrech	Trossteenbreek
Saxifrage, Prost's*	Saxifraga pedemontana prosti	Prost's-Steinbrech*	Prost's steenbreek
Saxifrage, Starry	Saxifraga stellaris	Stern-Steinbrech	Stersteenbreek
Scabious, Small	Scabiosa columbaria	Tauben-Skabiose	Duifkruid
Scorpion-vetch, Dwarf	Coronilla minima	Kleine Kronwicke	Klein kroonkruid
Scrub-rockrose, Alison*	Halimium alyssoides	Steinkraut-Strandröschen*	Schildzaadstrui-zonneroosje*
Scrub-rockrose, White*	Halimium umbellatum	Weisses Strandröschen*	Wit struikzonneroosje*
Self-heal, Large-flowered	Prunella grandiflora	Grosse Brunelle	Grote brunel
Senna, Bladder	Colutea arborescens	Blasenstrauch	Europese blazenstruik
Sermountain, Common	Laserpitium siler	Berg-Laserkraut	Smalbladig laserkruid
Sermountain, French	Laserpitium gallicum	Französisches Laserkraut	Frans laserkruid*
Sheep's-bit, Perrenial	Jasione laevis	Ausdauerndes Sand-glöckchen	Meerjarig zandblauwtje*
Smilax, Common	Smilax aspera	Stechwinde	Steekwinde
Smoke-tree	Cotinus coggygria	Perückenstrauch	Pruikenboom
Snakeshead, Pyrenean	Fritillaria pyrenaica	Pyrenäen-Schachblume	Pyreneeen kievitsbloem*
Snapdragon	Antirrhinum majus	Garten-Löwenmaul	Grote leeuwenbek
Snapdragon, Creeping	Asarina procumbens	Nierenblättriges Löwen-	Kruipende leeuwenbek

		maul	
Soapwort, Rock	Saponaria ocymoides	Rotes Seifenkraut	Rotszeepkruid
Solomon's-seal, Angular	Polygonatum odoratum	Wohlriechende Weisswurz	Welriekende salomonszegel
Solomon's-seal, Whorl-leaved	Polygonatum verticillatum	Quirlblättrige Weisswurz	Kranssalomonszegel
Sow-thistle, Plumer's*	Cicerbita plumieri	Französischer Milchlattich	Franse bergsla*
Speedwell, Prostrate	Veronica prostrata	Niederliegender Ehrenpreis	Liggende ereprijs
Spignel	Meum athamanthicum	Bärwurz	Bergvenkel
Spleenwort, Forez	Asplenium foreziense	Französischer Streifenfarn	Forez-streepvaren
Spleenwort, Forked	Asplenium septentrionale	Nördlicher Streifenfarn	Noordse streepvaren
Spleenwort, Maidenhair	Asplenium trichomanes	Braunstieliger Streifenfarn	Steenbreekvaren
Spruce	Picea sp.	Fichte	Spar
Spurge, Mediterranean	Euphorbia characias	Palisaden-Wolfsmilch	Zwartbloemige wolfsmelk*
Spurge, Nice	Euphorbia nicaeensis	Nizza-Wolfsmilch	Nice wolfsmelk*
Spurge, Steppe*	Euphorbia segueiriana	Steppen-Wolfsmilch	Zandwolfsmelk
Squill, Alpine	Scilla bifolia	Zweiblättriger Blaustern	Vroege sterhyacint
Squill, Autumn	Scilla autumnalis	Herbst-Blaustern	Herfststerhyacint
Squinancywort	Asperula cynanchica	Hügel-Meister	Kalkbedstro
Staehelina	Staehelina dubia	Zweifelhafte Strauchscharte	Staehelina*
Star-of-Bethlehem	Ornithogalum umbellatum	Dolden-Milchstern	Gewone vogelmelk
Star-of-Bethlehem, Early	Gagea bohemica	Felsen-Gelbstern	Vroege geelster*
Stitchwort, Wood	Stellaria nemorum	Hain-Sternmiere	Bosmuur
Stonecrop, Biting	Sedum acre	Scharfer Mauerpfeffer	Muurpeper
Stonecrop, Pale	Sedum sediforme	Felsen-Fetthenne	Rotsvetkruid*
Stonecrop, Pink	Sedum cepaea	Rispen-Mauerpfeffer	Omgebogen vetkruid
Stonecrop, Short-leaved*	Sedum brevifolium	Kurzblättriger Mauerpfeffer	Kortbladig vetkruid*
Stonecrop, Thick-leaved*	Sedum dasyphyllum	Buckliger Mauerpfeffer	Dik vetkruid
Stonecrope, Hoary*	Sedum hirsutum	Behaarter Mauerpfeffer	Harig vetkruid*
Stonecrope, White	Sedum album	Weisser Mauerpfeffer	Wit vetkruid
Storkbill, Wood	Geranium sylvaticum	Wald-Storchschnabel	Bosooievaarsbek
Succory, Lamb's	Arnoseris minima	Lämmersalat	Korensla
Sundew, Round-leaved	Drosera rotundifolia	Rundblättriger Sonnentau	Ronde zonnendauw
Swallow-wort	Vincetoxicum hirundinaria	Weisse Schwalbenwurz	Witte engbloem
Thistle, Carline	Carlina vulgaris	Gemeine Eberwurz	Driedistel
Thistle, Dwarf	Cirsium acaule	Stengellose Kratzdistel	Aarddistel
Thistle, Globe	Echinops sphaerocephalus	Drüsenblättrige Kugeldistel	Beklierde kogeldistel

Thistle, Woolly	Cirsium eriophorum	Wollige Kratzdistel	Wollige distel
Thrift, Girard's*	Armeria girardii	Girard's Grasnelke*	Girards engels gras*
Thyme, Cevennes*	Thymus nitens cebennis	Cevennes Thymian*	Cevennes tijm*
Toadflax, Daisy-leaved	Anarrhinum bellidifolium	Lochschlund	Madelief leeuwenbek*
Toadflax, Malling	Chaenorrhinum origani-folium	Origanblättriges Zwerglein-kraut	Marjoleinbekje
Toadflax, Pale	Linaria repens	Gestreiftes Leinkraut	Gestreepte leeuwenbek
Toadflax, Prostrate	Linaria supina	Niederliegendes Leinkraut	Liggende leeuwenbek
Tree, Mastic	Pistacia lentiscus	Mastixstrauch	Mastiekstruik
Tree, Olive	Olea europea	Ölbaum	Wilde olijfboom
Tree, Strawberry	Arbutus unedo	Erdbeerbaum	Aardbeiboom
Tree, Turpentine	Pistacia terebinthus	Terpentin-Pistazie	Terpentijnboom
Tree, Wayfaring	Viburnum lantana	Wolliger Schneeball	Wollige sneeuwbal
Trefoil, Red	Trifolium rubens	Purpur-Klee	Purperen klaver
Tulip, Wild	Tulipa australis	Südliche Tulpe	Zuidelijke tulp
Twayblade, Common	Listera ovata	Grosses Zweiblatt	Grote keverorchis
Twayblade, Lesser	Listera cordata	Kleines Zweiblatt	Kleine keverorchis
Valerian, Lecoque's Red*	Centranthus lecoqii	Lecoque's Spornblume*	Lecoque's spoorbloem*
Valerian, Marsh	Valeriana dioica	Kleiner Baldrian	Kleine valeriaan
Valerian, Three-leaved	Valeriana tripteris	Dreischnittiger Baldrian	Driebladige valeriaan*
Valerian, Tuberous*	Valeriana tuberosa	Knolliger Baldrian	Knolvaleriaan*
Venus' Looking-glass	Legousia speculum-veneris	Echter Frauenspiegel	Groot spiegelklokje
Vetch, False	Astragalus monspessulanus	Französischer Tragant	Montpellier hokjespeul
Vetch, Horse-shoe	Hippocrepis comosa	Gewöhnliche Hufeisenklee	Paardenhoefklaver
Violet, Early Dog	Viola reichenbachiana	Wald-Veilchen	Donkersporig bosviooltje
Violet, Heath Dog	Viola canina	Hunds-Veilchen	Hondsviooltje
Violet, Marsh	Viola palustris	Sumpf-Veilchen	Moerasviooltje
Viper's-grass, Hairy*	Scorzonera hirsuta	Behaartes Schwarzwurzel*	Behaarde schorseneer*
Viper's-grass, Purple	Scorzonera purpurea	Purper-Schwarzwurzel	Paarse schorseneer*
Weld	Reseda luteola	Färber-Wau	Wouw
Whin, Petty	Genista anglica	Englischer Ginster	Stekelbrem
Whitebeam	Sorbus aria	Echte Mehlbeere	Meelbes
Whitlowgrass, Yellow	Draba aizoides	Immergrünes Felsen-blümchen	Geel hongerbloempje
Willowherb, Alpine	Epilobium anagallidifolium	Gauchheilblättriges Weidenröschen	Alpenbasterwederik
Wintergreen, Green	Pyrola chlorantha	Grünblütiges Wintergrün	Groen wintergroen
Wintergreen, One-flowered	Moneses uniflora	Moosauge	Eenbloemig wintergroen
Wintergreen, Round-leaved	Pyrola rotundifolia	Rundblättriges Wintergrün	Rond wintergroen
Wolfsbane	Aconitum vulparia	Gelber Eisenhut	Gele monnikskap

SPECIES LIST & TRANSLATION

Woundwort, Hedge	Stachys sylvatica	Wald-Ziest	Bosandoorn
Woundwort, Perennial Yellow	Stachys recta	Aufrechte Ziest	Bergandoorn
Yellow-rattle, Lesser	Rhinanthus minor	Kleine Klappertopf	Kleine ratelaar
Yellow-rattle, Mediterranean	Rhinanthus mediterraneus	Niedriger Klappertopf	Mediterrane ratelaar*
Yellow-wort	Blackstonia perfoliata	Durchwachsener Bitterling	Zomerbitterling

Mammals

English	Scientific	German	Dutch
Bat, Daubenton's	Myotis daubentonii	Wasserfledermaus	Watervleermuis
Bat, Greater Horseshoe	Rhinolophus ferrum-equinum	Grosse Hufeisennase	Grote hoefijzerneus
Bat, Greater Mouse-eared	Myotis myotis	Grosses Mausohr	Vale vleermuis
Bat, Grey Long-eared	Plecotus austriacus	Graues Langohr	Grijze grootoorvleermuis
Bat, Leisler's	Nyctalus leisleri	Kleiner Abendsegler	Bosvleermuis
Bat, Lesser Horseshoe	Rhinolophus hipposideros	Kleine Hufeisennase	Kleine hoefijzerneus
Bat, Lesser Mouse-eared	Myotis blythii	Kleines Mausohr	Kleine vale vleermuis
Bat, Long-eared	Plecotus auritus	Braunes Langohr	Gewone grootoor-vleermuis
Bat, Natterer's	Myotis nattereri	Fransenfledermaus	Franjestaart
Bat, Notch-eared	Myotis emarginatus	Wimperfledermaus	Ingekorven vleermuis
Bat, Whiskered	Myotis mystacinus	Kleine Bartfledermaus	Baardvleermuis
Bear, Brown	Ursus arctos	Braunbär	Bruine beer
Beaver	Castor fiber	Bieber	Bever
Boar, Wild	Sus scrofa	Wildschwein	Wild zwijn
Deer, Fallow	Damus damus	Dammhirsch	Damhert
Deer, Red	Cervus elaphus	Rothirsch	Edelhert
Deer, Roe	Capreolus capreolus	Reh	Ree
Dormouse, Edible	Glis glis	Siebenschläfer	Relmuis
Ermine	Mustela erminea	Hermelin	Hermelijn
Fox	Vulpes vulpes	Rotfuchs	Vos
Genet	Genetta genetta	Ginsterkatze	Genetkat
Hare	Lepus europaeus	Hase	Haas
Hedgehog	Erinaceus europeus	Braubbustigel	Egel
Horse, Przewalski	Equus ferus przewalskii	Przewalski-Pferd	Przewalskipaard
Lynx	Lynx lynx	Luchs	Lynx
Marten, Pine	Martes martes	Baummarder	Boommarter
Marten, Stone	Martes foina	Steinmarder	Steenmarter

Mouflon	Ovis orientalis musimon	Mufflon	Moeflon
Pipistrelle, Common	Pipistrellus pipistrellus	Zwergfledermaus	Gewone dwergvleermuis
Pipistrelle, Kuhl's	Pipistrellus kuhlii	Weissrandfledermaus	Kuhl's dwergvleermuis
Pipistrelle, Savi's	Pipistrellus savii	Alpenfledermaus	Savi's dwergvleermuis
Pole-cat	Mustela putorius	Waldiltis	Bunzing
Rabbit	Oryctolagus cuniculus	Wildkaninchen	Konijn
Shrews	Soricidae	Spitzmäuse	Spitsmuizen
Squirrel, Red	Sciurus vulgaris	Eichhörnchen	Gewone eekhoorn
Stoat	Mustela ermina	Hermelin	Hermelijn
Weasel	Mustela nivalis	Mauswiesel	Wezel
Wolf	Canis lupus	Wolf	Wolf

Birds

English	Scientific	German	Dutch
Accentor, Alpine	Prunella collaris	Alpenbraunelle	Alpenheggemus
Accentor, Hedge	Prunella modularis	Heckenbraunelle	Heggenmus
Bee-eater	Merops apiaster	Bienenfresser	Bijeneter
Blackbird	Turdus merula	Amsel	Merel
Blackcap	Sylvia atricapilla	Mönchsgrasmücke	Zwartkop
Bullfinch	Pyrrhula pyrrhula	Gimpel	Goudvink
Bunting, Cirl	Emberiza cirlus	Zaunammer	Cirlgors
Bunting, Corn	Miliaria calandra	Grauammer	Grauwe gors
Bunting, Ortolan	Emberiza hortulana	Ortolan	Ortolaan
Bunting, Rock	Emberiza cia	Zippammer	Grijze gors
Bustard, Little	Tetrax tetrax	Zwergtrappe	Kleine trap
Buzzard	Buteo buteo	Mäusebussard	Buizerd
Buzzard, Honey	Pernis apivorus	Wespenbussard	Wespendief
Capercaillie	Tetrao urogallus	Auerhuhn	Auerhoen
Chaffinch	Fringilla coelebs	Buchfink	Vink
Chough, Red-billed	Pyrrhocorax pyrrhocorax	Alpenkrähe	Alpenkraai
Crossbill, Common	Loxia curvirostra	Fichtenkreuzschnabel	Kruisbek
Crow, Carrion	Corvus corone	Aaskrähe	Zwarte kraai
Cuckoo	Cuculus canorus	Kuckuck	Koekoek
Curlew, Stone	Burhinus oedicnemus	Triel	Griel
Dipper	Cinclus cinclus	Wasseramsel	Waterspreeuw
Dotterel	Charadrius morinellus	Mornellregenpfeifer	Morinelplevier
Dove, Collared	Streptopelia decaocto	Türkentaube	Turkse tortel
Dove, Stock	Columba oenas	Hohltaube	Holenduif
Dove, Turtle	Streptopelia turtur	Turteltaube	Tortelduif

Dunnock	*see Accentor, Hedge*		
Eagle, Golden	Aquila chrysaetos	Steinadler	Steenarend
Eagle, Short-toed	Circaetus gallicus	Schlangenadler	Slangenarend
Egret, Little	Egretta garzetta	Seidenreiher	Kleine zilverreiger
Finch, Citril	Serinus citrinella	Zitronengirlitz	Citroenkanarie
Firecrest	Regulus ignicapillus	Sommergoldhähnchen	Vuurgoudhaantje
Flycatcher, Pied	Ficedula hypoleuca	Trauerschnäpper	Bonte vliegenvanger
Flycatcher, Spotted	Muscicapa striata	Grauschnäpper	Grauwe vliegenvanger
Goldcrest	Regulus regulus	Wintergoldhähnchen	Goudhaan
Goldfinch	Carduelis carduelis	Distelfink	Putter
Goshawk	Accipiter gentilis	Habicht	Havik
Greenfinch	Carduelis chloris	Grünling	Groenling
Harrier, Hen	Circus cyaneus	Kornweihe	Blauwe kiekendief
Harrier, Montagu's	Circus pygargus	Wiesenweihe	Grauwe kiekendief
Hawfinch	Coccothraustes cocco-thraustes	Kernbeisser	Appelvink
Heron, Grey	Ardea cinerea	Graureiher	Blauwe reiger
Hobby	Falco subbuteo	Baumfalke	Boomvalk
Hoopoe	Upupa epops	Wiedehopf	Hop
Jackdaw	Corvus monedula	Dohle	Kauw
Jay	Garrulus glandarius	Eichelhäher	Gaai
Kestrel	Falco tinnunculus	Turmfalke	Torenvalk
Kingfisher	Alcedo atthis	Eisvogel	IJsvogel
Kite, Black	Milvus migrans	Schwarzmilan	Zwarte wouw
Kite, Red	Milvus milvus	Rotmilan	Rode wouw
Lark, Short-toed	Calandrella brachydactyla	Kurzzehenlerche	Kortteenleeuwerik
Linnet	Carduelis cannabina	Bluthänfling	Kneu
Magpie	Pica pica	Elster	Ekster
Mallard	Anas platyrhynchos	Stockente	Wilde eend
Martin, Crag	Ptyonoprogne rupestris	Felsenschwalbe	Rotszwaluw
Martin, House	Delichon urbica	Mehlschwalbe	Huiszwaluw
Merlin	Falco columbarius	Merlin	Smelleken
Nightingale	Luscinia megarhynchos	Nachtigall	Nachtegaal
Nightjar	Caprimulgus europaeus	Ziegenmelker	Nachtzwaluw
Nuthatch	Sitta europaea	Kleiber	Boomklever
Oriole, Golden	Oriolus oriolus	Pirol	Wielewaal
Ouzel, Ring	Turdus torquatus	Ring ouzel	Beflijster
Owl, Barn	Tyto alba	Schleiereule	Kerkuil
Owl, Eagle	Bubo bubo	Uhu	Oehoe
Owl, Little	Athene noctua	Steinkauz	Steenuil

Owl, Long-eared	Asio otus	Waldohreule	Ransuil
Owl, Scops	Otus scops	Zwergohreule	Dwergooruil
Owl, Tawny	Strix aluco	Waldkauz	Bosuil
Owl, Tengmalm's	Aegolius funereus	Raufusskauz	Ruigpootuil
Partridge, Grey	Perdix perdix	Rebhuhn	Patrijs
Partridge, Red-legged	Alectoris rufa	Rothuhn	Rode patrijs
Peregrine	Falco peregrinus	Wanderfalke	Slechtvalk
Pheasant	Phasianus colchicus	Fasan	Fazant
Pigeon, Feral	Columba livia f. domestica	Stadttaube	Stadsduif
Pigeon, Wood	Columba palumbus	Ringeltaube	Houtduif
Pipit, Meadow	Anthus pratensis	Wiesenpieper	Graspieper
Pipit, Tawny	Anthus campestris	Brachpieper	Duinpieper
Pipit, Tree	Anthus trivialis	Baumpieper	Boompieper
Pipit, Water	Anthus spinoletta	Bergpieper	Waterpieper
Quail	Coturnix coturnix	Wachtel	Kwartel
Raven	Corvus corax	Kolkrabe	Raaf
Redshank	Tringa totanus	Rotschenkel	Tureluur
Redstart, Black	Phoenicurus ochruros	Hausrotschwanz	Zwarte roodstaart
Redstart, Common	Phoenicurus phoenicurus	Gartenrotschwanz	Gekraagde roodstaart
Rock Thrush	Monticola saxatilis	Steinrötel	Rode rotslijster
Rock Thrush, Blue	Monticola solitarius	Blaumerle	Blauwe rotslijster
Roller	Coracias garrulus	Blauracke	Scharrelaar
Sandpiper, Common	Actitis hypoleucos	Flussuferläufer	Oeverloper
Sandpiper, Green	Tringa ochropus	Waldwasserläufer	Witgat
Serin	Serinus serinus	Girlitz	Europese kanarie
Shrike, Great Grey	Lanius excubitor	Raubwürger	Klapekster
Shrike, Red-backed	Lanius collurio	Neuntöter	Grauwe klauwier
Shrike, Southern Grey	Lanius meridionalis	Südlicher Raubwürger	Zuidelijke klapekster
Shrike, Woodchat	Lanius senator	Rotkopfwürger	Roodkopklauwier
Siskin	Carduelis spinus	Erlenzeisig	Sijs
Skylark	Alauda arvensis	Feldlerche	Veldleeuwerik
Snipe, Common	Gallinago gallinago	Bekassine	Watersnip
Sparrow, House	Passer domesticus	Haussperling	Huismus
Sparrow, Rock	Petronia petronia	Steinsperling	Rotsmus
Sparrow, Tree	Passer montanus	Feldsperling	Passer montanus
Sparrowhawk	Accipiter nisus	Sperber	Sperwer
Stilt, Black-winged	Himantopus himantopus	Stelzenläufer	Steltkluut
Stonechat	Saxicola torquata	Schwarzkehlchen	Roodborsttapuit
Swallow, Barn	Hirundo rustica	Rauchschwalbe	Boerenzwaluw
Swift, Alpine	Apus melba	Alpensegler	Alpengierzwaluw

SPECIES LIST & TRANSLATION

Swift, Common	Apus apus	Mauersegler	Gierzwaluw
Teal	Anas crecca	Krickente	Wintertaling
Thrush, Mistle	Turdus viscivorus	Misteldrossel	Grote lijster
Thrush, Song	Turdus philomelos	Singdrossel	Zanglijster
Tit, Blue	Parus caerulus	Blaumeise	Pimpelmees
Tit, Coal	Parus ater	Tannenmeise	Zwarte mees
Tit, Crested	Parus cristatus	Haubenmeise	Kuifmees
Tit, Great	Parus major	Kohlmeise	Koolmees
Tit, Long-tailed	Aegithalos caudatus	Schwanzmeise	Staartmees
Tit, Marsh	Parus palustris	Sumpfmeise	Glanskop
Treecreeper, Eurasian	Certhia familiaris	Waldbaumläufer	Taigaboomkruiper
Treecreeper, Short-toed	Certhia brachydactyla	Gartenbaumläufer	Boomkruiper
Vulture, Black	Aegypius monachus	Mönchsgeier	Monniksgier
Vulture, Egyptian	Neophron percnopterus	Schmutzgeier	Aasgier
Vulture, Griffon	Gyps fulvus	Gänsegeier	Vale gier
Wagtail, Blue-headed	Motacilla flava flava	Wiesenschafstelze	Gele kwikstaart
Wagtail, Grey	Motacilla cinerea	Gebirgsstelze	Grote gele kwikstaart
Wagtail, White	Motacilla alba	Bachstelze	Witte kwikstaart
Wallcreeper	Tichodroma muraria	Mauerläufer	Rotskruiper
Warbler, Bonelli's	Phylloscopus bonelli	Berglaubsänger	Bergfluiter
Warbler, Cetti's	Cettia cetti	Seidensänger	Cetti's zanger
Warbler, Dartford	Sylvia undata	Provencegrasmücke	Provençaalse grasmus
Warbler, Garden	Sylvia borin	Gartengrasmücke	Tuinfluiter
Warbler, Melodious	Hippolais polyglotta	Orpheusspötter	Orpheusspotvogel
Warbler, Orphean	Sylvia hortensis	Orpheusgrasmücke	Orpheusgrasmus
Warbler, Sardinian	Sylvia melanocephala	Samtkopf-Grasmücke	Kleine zwartkop
Warbler, Subalpine	Sylvia cantillans	Weissbart-Grasmücke	Baardgrasmus
Wheatear, Black-eared	Oenanthe hispanica	Mittelmeer-Steinschmätzer	Blonde tapuit
Wheatear, Northern	Oenanthe oenanthe	Steinschmätzer	Tapuit
Whinchat	Saxicola rubetra	Braunkehlchen	Paapje
Whitethroat	Sylvia communis	Dorngrasmücke	Grasmus
Woodcock	Scolopax rusticola	Waldschnepfe	Houtsnip
Woodlark	Lullula arborea	Heidelerche	Boomleeuwerik
Woodpecker, Black	Dryocopus martius	Schwarzspecht	Zwarte Specht
Woodpecker, Great Spotted	Dendrocopos major	Buntspecht	Grote bonte specht
Woodpecker, Green	Picus viridis	Grünspecht	Groene specht
Woodpecker, Lesser Spotted	Dendrocopos minor	Kleinspecht	Kleine bonte specht
Wryneck	Jynx torquilla	Wendehals	Draaihals
Yellowhammer	Emberiza citrinella	Goldammer	Geelgors

Reptiles and Amphibians

English	Scientific	German	Dutch
Adder	Vipera berus	Kreuzotter	Adder
Frog, Graf's Hybrid	Rana grafi	Grafscher Hybridfrosch	Graf's bastaardkikker
Frog, Grass	Rana temporaria	Grasfrosch	Bruine kikker
Frog, Iberian Water	Rana perezi	Iberischer Wasserfrosch	Iberische groene kikker
Frog, Marsh	Rana ridibunda	Seefrosch	Meerkikker
Frog, Parsley	Pelodytes punctatus	Westlicher Schlammtaucher	Groengestipte kikker
Frog, Stripeless Tree	Hyla meridionalis	Mittelmeer-Laubfrosch	Mediterrane boomkikker
Lizard, Green	Lacerta virides	Smaragdeidechse	Oostelijke smaragdhagedis
Lizard, Iberian Wall	Podarcis hispanica	Spanische Mauereidechse	Spaanse muurhagedis
Lizard, Ocellated	Lacerta lepida	Perleidechse	Parelhagedis
Lizard, Sand	Lacerta agilis	Zauneidechse	Zandhagedis
Lizard, Viviparous	Lacerta vivipara	Bergeidechse	Levendbarende hagedis
Lizard, Wall	Podarcis muralis	Mauereidechse	Muurhagedis
Lizard, Western Green	Lacerta bilineata	Westliche Smaragdeidechse	Westelijke smaragdhagedis
Newt, Palmate	Triturus helveticus	Fadenmolch	Vinpootsalamander
Salamander, Fire	Salamandra salamandra	Feuersalamander	Vuursalamander
Snake, Aesculapian	Elaphe longissima	Äskulapnatter	Esculaapslang
Snake, Grass	Natrix natrix	Ringelnatter	Ringslang
Snake, Ladder	Elaphe scalaris	Treppennatter	Trapslang
Snake, Montpellier	Malpolon monspessulanus	Eidechsennatter	Hagedisslang
Snake, Smooth	Coronella austriaca	Schlingnatter	Gladde slang
Snake, Southern Smooth	Coronella girondica	Girondische Glattnatter	Girondische gladde slang
Snake, Viperine	Natrix maura	Vipernatter	Adderringslang
Snake, Western Whip	Coluber viridiflavus	Gelbgrüne Zornnatter	Geelgroene toornslang
Terrapin, European Pond	Emys orbicularis	Europäische Sumpfschildkröte	Europese moerasschildpad
Toad, Common	Bufo bufo	Erdkröte	Gewone pad
Toad, Midwife	Alytes obstetricans	Geburtshelferkröte	Vroedmeesterpad
Toad, Natterjack	Bufo calamita	Kreuzkröte	Rugstreeppad
Toad, Yellow-bellied	Bombina variegata	Gelbbauchunke	Geelbuikvuurpad
Viper, Asp	Vipera aspis	Aspisviper	Aspisadder
Worm, Slow	Anguis fragilis	Blindschleiche	Hazelworm

Insects

English	Scientific	German	Dutch
Admiral, Red	Vanessa atalanta	Admiral	Atalanta
Admiral, Southern White	Limenitis reducta	Blauschwarzer Eisvogel	Blauwe ijsvogelvlinder
Admiral, White	Limenitis camilla	Kleiner Eisvogel	Kleine ijsvogelvlinder
Apollo	Parnassius apollo	Apollofalter	Apollovlinder
Apollo, Clouded	Parnassius mnemosyne	Schwarze Apollo	Zwarte apollovlinder
Argus, Brown	Aricia agestis	Kleiner Sonnenröschen-Bläuling	Bruin blauwtje
Argus, Mountain	Aricia artaxerxes	Grosser Sonnenröschen-Bläuling	Vals bruin blauwtje
Argus, Scotch	Erebia aethiops	Graubindiger Mohrenfalter	Zomererebia
Ascalaphid, Yellow-winged*	Libelloides coccajus	Libellen-Schmetterlings-haft	Gewone vlinderhaft
Ascalaphids	Ascalaphidae	Schmetterlingshafte	Vlinderhaften
Beauty, Camberwell	Nymphalis antiopa	Trauermantel	Rouwmantel
Beetle, Hoplia	Hoplia coerulea	Hoplia*	Hoplia*
Beetle, Longhorn	Cerambycidae	Bockkäfer	Boktorren
Beetle, Stag	Lucanus cervus	Hirschkäfer	Vliegend hert
Blue, (Western) Furry	Polyommatus dolus	Cevennes-Blauling*	Westelijk vachtblauwtje
Blue, Adonis	Polyommatus bellargus	Himmelblauer Bläuling	Adonisblauwtje
Blue, Amanda's	Polyommatus amandus	Vogelwicken-Bläuling	Wikkeblauwtje
Blue, Baton	Pseudophilotes baton	Graublauer Bläuling	Klein tijmblauwtje
Blue, Black-eyed	Glaucopsyche melanops	Schwarz-Auge Blauling*	Spaans bloemenblauwtje
Blue, Chalkhill	Polyommatus coridon	Silbergrüner Bläuling	Bleek blauwtje
Blue, Chapman's	Polyommatus thersites	Kleine Esparsetten -Bläuling	Esparcetteblauwtje
Blue, Chequered	Scolitantides orion	Fetthennen-Bläuling	Vetkruidblauwtje
Blue, Escher's	Polyommatus escheri	Escher-Bläuling	Groot tragantblauwtje
Blue, Green-underside	Glaucopsyche alexis	Himmelblauer Steinklee-bläuling	Bloemenblauwtje
Blue, Large	Maculinea arion	Schwarzgefleckten Bläuling	Tijmblauwtje
Blue, Little	Cupido minimus	Zwerg-Bläuling	Dwergblauwtje
Blue, Long-tailed	Lampides boeticus	Grosser Wander-Bläuling	Tijgerblauwtje
Blue, Meleager's	Polyommatus daphnis	Zahnflügel-Bläuling	Getand blauwtje
Blue, Mountain Alcon	Maculinea rebeli	Kreuzenzian-Ameisen-Bläuling	Berggentiaanblauwtje
Blue, Osiris	Cupido osiris	Kleiner Alpenbläuling	Zuidelijk dwergblauwtje
Blue, Provençal Short-tailed	Cupido alcetas	Südlicher Kurz-geschwänzter Bläuling	Zuidelijk staartblauwtje

Blue, Ripart's Anomalous	Polyommatus ripartii	Südliche Esparsetten -Bläuling*	Zuidelijk esparcette- blauwtje
Bluet, Azure	Coenagrion puella	Hufeisen-Azurjungfer	Azuurwaterjuffer
Bluet, Dainty	Coenagrion scitulum	Gabel-Azurjungfer	Gaffelwaterjuffer
Bluet, Variable	Coenagrion pulchellum	Fledermaus-Azurjungfer	Variabele waterjuffer
Bluetail, Small	Ischnura pumilio	Kleine Pechlibelle	Tengere grasjuffer
Brown, Arran	Erebia ligea	Weissbindiger Mohrenfalter	Boserebia
Brown, Wall	Lasiommata megera	Mauerfuchs	Argusvlinder
Bush-cricket, Southern Saw-tailed	Barbitistes fischeri	Südfranzösische Säbelschrecke	Zuidelijke zaagsprinkhaan*
Bush-cricket, Saddle-back	Ephippiger ephippiger	Steppen-Sattelschrecke	Zadelsprinkhaan
Butterfly, Nettle-tree	Libythea celtis	Zürgelbaum-Schnauzen- falter	Snuitvlinder
Cardinal	Argynnis pandora	Kardinal	Kardinaalsmantel
Chafer, Rose	Cetonia aurata	Gemeiner Rosenkäfer	Gouden tor
Chaser, Broad-bodied	Libellula depressa	Plattbauch	Platbuik
Chaser, Four-spotted	Libellula quadrimaculata	Vierfleck	Viervlek
Chaser, Splendid	Macromia splendens	Europäischer Flussherrscher	Prachtlibel
Cicada	Cicada orni	Mannazikade	Kleine zangcicade*
Cleopatra	Gonepteryx cleopatra	Mittelmeer-Zitronenfalter	Cleopatra
Clubtail, Pronged	Gomphus graslinii	Französische Keiljungfer	Gevorkte rombout
Comma	Polygonia c-album	C-Falter	Gehakkelde aurelia
Copper, Purple-edged	Lycaena hippothoe	Lilagold-Feuerfalter	Rode vuurvlinder
Copper, Purple-shot	Lycaena alciphron	Violetter Feuerfalter	Violette vuurvlinder
Copper, Scarce	Lycaena virgaureae	Dukatenfalter	Morgenrood
Copper, Sooty	Lycaena tityrus	Brauner Feuerfalter	Bruine vuurvlinder
Damselfly, Large Red	Pyrrhosoma nymphula	Frühe Adonisjungfer	Vuurjuffer
Darter, Black	Sympetrum danae	Schwarze Heidelibelle	Zwarte heidelibel
Darter, Common	Sympetrum striolatum	Grosse Heidelibelle	Bruinrode heidelibel
Darter, Yellow-winged	Sympetrum flaveolum	Gefleckte Heidelibelle	Geelvlekheidelibel
Demoiselle, Copper	Calopteryx haemorrhoidalis	Rote Prachtlibelle	Koperen beekjuffer
Demoiselle, Western	Calopteryx xanthostoma	Südwestliche Prachtlibelle	Iberische beekjuffer
Emerald, Downy	Cordulia aenea	Gemeine Smaragdlibelle	Smaragdlibel
Emperor, Blue	Anax imperator	Grosse Königslibelle	Grote keizerlibel
Emperor, Lesser Purple	Apatura ilia	Kleiner Schillerfalter	Kleine weerschijnvlinder
Emperor, Purple	Apatura iris	Grosser Schillerfalter	Grote weerschijnvlinder
Featherleg, Blue	Platycnemis pennipes	Blaue Federlibelle	Blauwe breedscheenjuffer
Featherleg, White	Platycnemis latipes	Weisse Federlibelle	Witte breedscheenjuffer
Fly, Bee	Bombylius major	Grosser Wollschweber	Gewone wolzwever
Fly, Caddis	Trichoptera	Köcherfliegen	Schietmotten

Fritillary, Duke of Burgundy	Hamearis lucina	Schlüsselblumen-Würfelfalter	Sleutelbloemvlinder
Fritillary, False Heath	Melitaea diamina	Baldrian-Scheckenfalter	Woudparelmoervlinder
Fritillary, Glanville	Melitaea cinxia	Wegerich-Scheckenfalter	Veldparelmoervlinder
Fritillary, Heath	Melitaea athalia	Gemeine Scheckenfalter	Bosparelmoervlinder
Fritillary, High Brown	Argynnis adippe	Feuriger Perlmutterfalter	Bosrandparelmoervlinder
Fritillary, Knapweed	Melitaea phoebe	Flockenblumen-Scheckenfalter	Knoopkruidparelmoervlinder
Fritillary, Lesser Marbled	Brenthis ino	Mädesüss-Perlmutterfalter	Purperstreep-parelmoervlinder
Fritillary, Marbled	Brenthis daphne	Brombeer-Perlmuttfalter	Braamparelmoervlinder
Fritillary, Marsh	Eurodryas aurinia	Skabiosen-Scheckenfalter	Moerasparelmoervlinder
Fritillary, Meadow	Melitaea parthenoides	Westlicher Scheckenfalter	Westelijke parelmoervlinder
Fritillary, Niobe	Argynnis niobe	Stiefmütterchen-Perlmutterfalter	Duinparelmoervlinder
Fritillary, Pearl-bordered	Boloria euphrosyne	Frühlings-Perlmuttfalter	Zilvervlek
Fritillary, Provençal	Melittaea deione	Leinkraut-Scheckenfalter	Spaanse parelmoervlinder
Fritillary, Queen of Spain	Issoria lathonia	Kleiner Perlmutterfalter	Kleine parelmoervlinder
Fritillary, Silver-washed	Argynnis paphia	Kaisermantel	Keizersmantel
Fritillary, Small Pearl-bordered	Boloria selene	Braunfleckiger Perlmutterfalter	Zilveren maan
Fritillary, Spotted	Melitaea didyma	Roter Scheckenfalter	Tweekleurige parelmoervlinder
Fritillary, Titania's	Boloria titania	Natterwurz-Perlmutterfalter	Titania's parelmoervlinder
Fritillary, Weaver's	Boloria dia	Magerrasen-Perlmutterfalter	Akkerparelmoervlinder
Gatekeeper, Common	Pyronia tithonus	Rotbraunes Ochsenauge	Oranje zandoogje
Gatekeeper, Southern	Pyronia cecilia	Südliches Ochsenauge	Zuidelijk oranje zandoogje
Gatekeeper, Spanish	Pyronia bathseba	Spanischer Ochsenauge	Spaans oranje zandoogje
Goldenring, Common	Cordulegaster boltonii	Zweigestreifte Quelljungfer	Gewone bronlibel
Goldenring, Sombre	Cordulegaster bidentata	Gestreifte Quelljungfer	Zuidelijke bronlibel
Grayling	Hipparchia semele	Ockerbindige Samtfalter	Heivlinder
Grayling, False	Arethusana arethusa	Rotbindiger Samtfalter	Oranje steppevlinder
Grayling, Great Banded	Brintesia circe	Weisser Waldportier	Witbandzandoog
Grayling, Woodland	Hipparchia fagi	Grosser Waldportier	Grote boswachter
Hairstreak, Blue-spot	Satyrium spini	Kreuzdorn-Zipfelfalter	Wegedoornpage
Hairstreak, False Ilex	Satyrium esculi	Südlicher Eichen-Zipfelfalter	Spaanse eikenpage

Hairstreak, Green	Callophrys rubi	Grüner Zipfelfalter	Groentje
Hairstreak, Ilex	Satyrium ilicis	Brauner Eichen-Zipfelfalter	Bruine eikenpage
Hairstreak, Sloe	Satyrium acaciae	Kleiner Schlehen-Zipfelfalter	Kleine sleedoornpage
Hairstreak, Spanish Purple	Laeosopis roboris	Spanischer Blauer Zipfelfalter	Essenpage
Hairstreak, White-letter	Satyrium w-album	Ulmen-Zipfelfalter	Iepenpage
Hawker, Blue	Aeshna cyanea	Blaugrüne Mosaikjungfer	Blauwe glazenmaker
Hawker, Moorland	Aeshna juncea	Torf-Mosaikjungfer	Venglazenmaker
Heath, Chestnut	Coenonympha glycerion	Rostbraunes Wiesenvögelchen	Roodstreep-hooibeestje
Heath, Dusky	Coenonympha dorus	Dorus Wiesenvögelchen	Bleek hooibeestje
Heath, Pearly	Coenonympha arcania	Weissbindiges Wiesenvögelchen	Tweekleurig hooibeestje
Heath, Small	Coenonympha pamphilus	Kleines Wiesenvögelchen	Hooibeestje
Hermit	Chazara briseis	Berghexe	Heremiet
Lady, Painted	Vanessa cardui	Distelfalter	Distelvlinder
Mantis, Hooded Praying*	Empusa pennata	Kapuze-Gottesanbeterin*	Kapbidsprinkhaan*
Mantis, Praying	Mantis religiosa	Gottesanbeterin	Bidsprinkhaan
Moth, Silk	Bombyx mori	Seidenspinner	Zijdevlinder
Orange-tip, Provence	Anthocharis euphenoides	Gelber Aurorafalter	Geel oranjetipje
Owlfly, European	Libelloides coccajus	Libellen-Schmetterlingshaft	Gewone vlinderhaft
Pasha, Two-tailed	Charaxes jasius	Erdbeerbaumfalter	Pasja
Peacock	Inachis io	Tagpfauenauge	Dagpauwoog
Pincertail, Great	Onychogomphus uncatus	Grosse Zangenlibelle	Grote tanglibel
Pincertail, Small	Onychogomphus forcipatus	Kleine Zangenlibelle	Kleine tanglibel
Ringlet, Autumn	Erebia neoridas	Herbst-Mohrenfalter*	Herfsterebia
Ringlet, Bright-eyed	Erebia oeme	Doppelaugen-Mohrenfalter	Bontoogerebia
Ringlet, Large	Erebia euryale	Weissbindige Bergwald-Mohrenfalter	Grote erebia
Ringlet, Mountain	Erebia epiphron	Knochs Mohrenfalter	Bergerebia
Ringlet, Ottoman Brassy	Erebia ottomana	Östlicher Mohrenfalter*	Oostelijke glanserebia
Ringlet, Piedmont	Erebia meolans	Gelbbindiger Mohrenfalter	Donkere erebia
Ringlet, Woodland	Erebia medusa	Rundaugen-Mohrenfalter	Voorjaarserebia
Satyr, Great Sooty	Satyrus ferula	Weisskernauge	Grote saterzandoog
Skimmer, Black-tailed	Orthetrum cancellatum	Grosser Blaupfeil	Gewone oeverlibel
Skimmer, Keeled	Orthetrum coerulescens	Kleiner Blaupfeil	Beekoeverlibel
Skimmer, Southern	Orthetrum brunneum	Südlicher Blaupfeil	Zuidelijke oeverlibel
Skimmer, White-tipped	Orthetrum albistylum	Östlicher Blaupfeil	Witpuntoeverlibel

SPECIES LIST & TRANSLATION

Skipper, Cinquefoil	Pyrgus cirsii	Spätsommer-Würfel-Dickkopffalter	Rood spikkeldikkopje
Skipper, Dingy	Erynnis tages	Kronwicken-Dickkopffalter	Bruin dikkopje
Skipper, Grizzled	Pyrgus malvae	Kleiner Würfel-Dickkopffalter	Aardbeivlinder
Skipper, Grizzled Foulquier's	Pyrgus foulquieri	Fouquier's Würfel-Dickkopffalter*	Frans spikkeldikkopje
Skipper, Grizzled Large	Pyrgus alveus	Sonnenröschen-Würfel-Dickkopffalter	Groot spikkeldikkopje
Skipper, Grizzled Oberthur's	Pyrgus armoricanus	Zweibrütiger Würfel-Dickkopffalter	Bretons spikkeldikkopje
Skipper, Grizzled Rosy	Pyrgus onopordi	Ambossfleck-Würfel-Dickkopffalter	Aambeeldspikkeldikkopje
Skipper, Large	Ochlodes faunus	Rostfarbiger Dickkopffalter	Groot dikkopje
Skipper, Olive	Pyrgus serratulae	Schwarzbrauner Würfelfalter	Voorjaarsspikkeldikkopje
Skipper, Safflower	Pyrgus carthami	Steppenheiden-Würfel-Dickkopffalter	Witgezoomd spikkeldikkopje
Skipper, Small	Thymelicus sylvestris	Braunkolbiger Braun-Dickkopffalter	Geelsprietdikkopje
Spectre, Western	Boyeria irene	Westliche Geisterlibelle	Schemerlibel
Spider, Gravel Wolf*	Arctosa cinerea	Flussuferwolfspinne	Grindwolfspin
Spider, Wasp	Argiope bruennichi	Wespenspinne	Wespspin
Spider, Wolf	See Tarantula		
Spreadwing, Robust	Lestes dryas	Glänzende Binsenjungfer	Tangpantserjuffer
Swallowtail, Common	Papilio machaon	Schwalbenschwanz	Koninginnenpage
Swallowtail, Scarce	Iphiclides podalirius	Segelfalter	Koningspage
Tarantula	Lycosa narbonensis	Tarantula	Tarantula
Tortoiseshell, Large	Nymphalis polychloros	Grosser Fuchs	Grote vos
White, Bath	Pontia daplidice	Reseda Falter	Resedawitje
White, Black-veined	Aporia crataegi	Baumweissling	Groot geaderd witje
White, Esper's Marbled	Melanargia russiae	Südliches Schachbrett	Zuidelijk dambordje
White, Marbled	Melanargia galathea	Schachbrett	Dambordje
White, Western Dappled	Euchloe crameri	Westlicher Gesprenkelter Weissling	Westelijk marmerwitje
White, Wood	Leptidea sinapis	Senfweissling	Boswitje
Whiteface, Small	Leucorrhinia dubia	Kleine Moosjungfer	Venwitsnuitlibel
Wood, Speckled	Pararge aegeria	Waldbrettspiel	Bont zandoogje
Yellow, Berger's Clouded	Colias alfacariensis	Hufeisenklee-Gelbling	Zuidelijke luzernevlinder
Yellow, Clouded	Colias crocea	Postillion	Oranje luzernevlinder